ENTROPY
The
TRUTH
The Whole
TRUTH
and Nothing but the
TRUTH

ENTROPY
The
TRUTH
The Whole
TRUTH
and Nothing but the
TRUTH

Arieh Ben-Naim

The Hebrew University of Jerusalem, Israel

World Scientific

NEW JERSEY · LONDON · SINGAPORE · BEIJING · SHANGHAI · HONG KONG · TAIPEI · CHENNAI · TOKYO

Published by

World Scientific Publishing Co. Pte. Ltd.
5 Toh Tuck Link, Singapore 596224
USA office: 27 Warren Street, Suite 401-402, Hackensack, NJ 07601
UK office: 57 Shelton Street, Covent Garden, London WC2H 9HE

Library of Congress Cataloging-in-Publication Data
Names: Ben-Naim, Arieh, 1934– author.
Title: Entropy : the truth, the whole truth, and nothing but the truth /
  Arieh Ben-Naim, The Hebrew University of Jerusalem, Israel.
Description: Hackensack, NJ : World Scientific, [2016] |
  Includes bibliographical references and index.
Identifiers: LCCN 2016041556| ISBN 9789813147669 (hardcover ; alk. paper) |
  ISBN 9813147660 (hardcover ; alk. paper) | ISBN 9789813147676 (pbk. ; alk. paper) |
  ISBN 9813147679 (pbk. ; alk. paper)
Subjects: LCSH: Entropy. | Second law of thermodynamics.
Classification: LCC QC318.E57 B4583 2016 | DDC 536/.73--dc23
LC record available at https://lccn.loc.gov/2016041556

British Library Cataloguing-in-Publication Data
A catalogue record for this book is available from the British Library.

Copyright © 2017 by World Scientific Publishing Co. Pte. Ltd.

All rights reserved. This book, or parts thereof, may not be reproduced in any form or by any means, electronic or mechanical, including photocopying, recording or any information storage and retrieval system now known or to be invented, without written permission from the publisher.

For photocopying of material in this volume, please pay a copying fee through the Copyright Clearance Center, Inc., 222 Rosewood Drive, Danvers, MA 01923, USA. In this case permission to photocopy is not required from the publisher.

Printed in Singapore by Mainland Press Pte Ltd.

Dedicated to my dear wife Ruby
who maintains my low entropy life
by feeding me with delicious negative entropy food.

"Meaningless! Meaningless!" says the Teacher.
"Utterly meaningless! Everything is meaningless."
Ecclesiastes (1,2)

הֲבֵל הֲבָלִים אָמַר קֹהֶלֶת, הֲבֵל הֲבָלִים, הַכֹּל הָבֶל (קהלת א,ב)

Contents

Preface   xiii

Acknowledgments   xix

Test Yourself Before Reading This Book   1

Chapter 1.   The Various Definitions of Entropy and the Second Law of Thermodynamics   5

1.1   Introduction   5
1.2   The Clausius Definition and Its Extensions   7
    1.2.1   Heat Engines and Carnot's Efficiency   7
    1.2.2   Clausius' Definition of Entropy   10
    1.2.3   Entropy Change in an Expansion of an Ideal Gas   17
1.3   The Microscopic Definition of Entropy   22
    1.3.1   Introduction   22
    1.3.2   Boltzmann's Definition of Entropy   23
    1.3.3   Objections to Boltzmann's Entropy   24
    1.3.4   Absolute Values of Entropy and the Third Law   25
    1.3.5   Extension of the Second Law to Non-Isolated Systems   26

1.4	The Definition of Entropy Based on the Shannon's Measure of Information	27
	1.4.1 Introduction	27
	1.4.2 What is the SMI?	29
	1.4.3 Uniform and Non-Uniform Distribution	32
	1.4.4 Interpretation of the SMI	41
	1.4.5 Applying the SMI to Obtain the Entropy of an Ideal Gas	45
	1.4.6 Extension to Systems with Interacting Particles	57
1.5	The Entropy Formulation of the Second Law of Thermodynamics	58
	1.5.1 An Expansion of an Ideal Gas in an Isolated System	59
	1.5.2 What Drives the System to an Equilibrium State?	61
	1.5.3 The Evolution of the SMI in the Expansion Process	67
	1.5.4 Summary of Facts	69
	1.5.5 Examples of Processes Associated with the Second Law	74
1.6	The Different Formulations of the Second Law of Thermodynamics	77
	1.6.1 A Brief Survey of the Various Ensembles	78
	1.6.2 The Formulation of the Second Law for Isolated Systems	83
	1.6.3 The (T, V, N) Formulation of the Second Law . .	93
	1.6.4 The (T, P, N) Formulation of the Second Law . .	94
1.7	A Few Misdefinitions of Entropy and the Second Law . .	96
1.8	Conclusion	98

Chapter 2.	Interpretation and Misinterpretations of Entropy	103
2.1	Introduction	103
2.2	What are the Sources of the Mystery?	105
	2.2.1 The Very New Word "Entropy"	105
	2.2.2 Entropy Always Increases	108
	2.2.3 Saying that Entropy is a Mystery, Renders Entropy a Mystery	109
	2.2.4 Entropy, as the Almighty "Driving Force" Underlying All Processes	109
	2.2.5 The Multiple Interpretations of Entropy	111
	2.2.6 The "Reversal Paradox" and the "Recurrence Paradox"	113
	2.2.7 The Unwarranted Application of Entropy to Life and the Entire Universe	115
2.3	A Necessary Criterion a Descriptor of Entropy Must Satisfy	115
	2.3.1 Volume Exchange Between Two Compartments	116
	2.3.2 Energy Exchange Between Two Compartments	118
	2.3.3 Material Exchange Between Two Compartments	119
2.4	The Association of Entropy with Disorder	120
2.5	The Association of Entropy with the Spreading/Dispersion/Sharing of Energy	128
2.6	The Association of Entropy with Information	135
2.7	Does Entropy Depend on Our Knowledge About the System?	140
	2.7.1 Does Entropy Depend on the Precision We Choose to Describe the Configuration of the System?	141

       2.7.2 Does Entropy Depend on Our Knowledge
              of the Intermolecular Interactions? 142
       2.7.3 Does Entropy Depend on the Knowledge
              of the Composition of the System? 144
2.8   Entropy as a Measure of Freedom 148
2.9   Entropy as a Measure of Possibilities 149
2.10 Entropy as a Measure of Irreversibility 150
2.11 Entropy as "Heat Loss," "Thermal Energy Not
      Available To Work," "Unavailable Energy," Etc. 152
2.12 A Challenging Problem 153
2.13 Caveat on the Interpretation of Entropy
      as Uncertainty 155
2.14 Conclusion 156

Chapter 3. Applications and Misapplications of Entropy 157

3.1   Introduction 157
3.2   Residual Entropy 160
      3.2.1 Residual Entropy of Ice 165
3.3   Application of the Second Law for Processes
      in an Isolated System 173
3.4   Applications of Entropy and the Second Law
      for a Constant Temperature and Constant
      Pressure System 174
      3.4.1 Formation of Hydrogen Molecules
             from Hydrogen Atoms 175
      3.4.2 Protein-Protein Association 178
      3.4.3 A Simple Chemical Equilibrium 181
      3.4.4 Protein Folding 183
3.5   Entropy Change in Phase Transitions 187
      3.5.1 Phase Diagram of Water 189
3.6   Trouton's Law 191
3.7   The Entropy of Solvation of Argon in Water 193
3.8   "Entropy of Mixing" 199

3.9   Racemization as a Pure Deassimilation Process 206
3.10 Misusing Entropy in Explaining the Low Solubility
      of Argon in Water 214
3.11 Application of Entropy and the Second Law
      to Living Systems 216
3.12 Application of Entropy and the Second Law
      to the Entire Universe 219
3.13 The Association of Entropy with the Arrow of Time . . 223
3.14 Conclusion 225

Test Yourself After Reading This Book   227

Notes   233

References and Suggested Reading   267

Index   273

Preface

Entropy is one of the most interesting and useful concepts in physics. It is also the most misused and sometimes also abused concept. I do not know of any other concept in the sciences which is *well-defined* on one hand, yet has aroused so much mystery, on the other hand.

I believe that much of the mystery enshrouding entropy is a result of the fact that for over a hundred years scientists have endeavored to find a simple, qualitative explanation for entropy. In doing so, many interpretations were suggested; none of them were proven to be correct. This caused great confusion among scientists who wrote about entropy, as well as among non-scientists who read what scientists wrote about entropy.

I believe that the seed of confusion can be attributed to Clausius, who introduced the concept of entropy, and perhaps innocently *overgeneralized* its applicability to the entire universe. Another over-application of entropy can be found in the writings of Schrödinger, Monod, Brillouin (1962), and many others associating entropy with life phenomena.

It is not surprising therefore to find authors of popular science books ascribing to entropy all kinds of super powers, explaining everything that is unexplainable. In my opinion, such practices only deepen and entrench the mystery associated with entropy.

Although I have written several books on entropy and the Second Law, I feel compelled to write this book in order to "clean up" the mess left behind by so many authors' writings on entropy and the Second Law.

This book is about the proper *definitions* of entropy, the valid *interpretations* of entropy, and some examples of useful *applications* of entropy and the Second Law.

Thanks to the Shannon's measure of information (SMI), we now have a crystal clear meaning of entropy. Ironically, Shannon himself, by naming his measure *entropy*, unintentionally contributed to the confusion about the meaning of entropy. I believe that once you grasp the meaning of entropy, the famous quotation from von Neumann that "no one knows what entropy really is," will no longer apply to you.

I also believe that once you "see the meaning of entropy through the lens of the SMI," you will be equipped with a powerful tool to critically assess the flood of fancy and pompous statements made by authors of popular science books on entropy and the Second Law.

Quite recently, two books expressing diametrically different views about entropy and the Second Law were published. In a nutshell, one claims that entropy is the driving force underlying *everything that happens*; that the Second Law governs any *process* in the universe. The second claims that entropy is *not needed*, that the Second Law is a *faulty logic*, and that entropy and the Second Law are *Science's biggest blunders*. In the notes to the preface I will quote the relevant paragraphs describing in more detail the contents of the books by Kafri and Kafri (2013),[1] and by Mayhew (2015a).[2]

I do not agree with either of these extreme views.[3] The main purpose of this book is to show that both entropy and the Second Law are meaningful and useful concepts, on one hand, and both have limitations in their applicability, on the other hand.

This book is organized into three chapters. The first chapter presents three definitions of entropy and the Second Law. The first is based on purely experimental quantities attributed to Clausius. The second is based on the number of *microstates* due to Boltzmann and Gibbs. The third is based on the Shannon's measure of information (SMI). The third, which also happens to be the more recent definition, has two parts. The first part was developed by Jaynes (1957a, 1957b, 1965) and Katz (1967) who used the so-called principle of Max Entropy to obtain the equilibrium distributions of statistical mechanics. The second part makes use of the Max SMI principle to obtain the equilibrium distribution of locations and velocities (or momenta) of all the particles in a (classical) macroscopic system, then uses these distributions to obtain the *entropy* of an ideal gas at equilibrium. This part was developed by Ben-Naim (2008). It also shows the relationship between the SMI and the probabilistic interpretation of the Second Law.

In my view, the third approach is both simpler and superior compared with the older approaches. It is also the only definition of entropy which provides a simple, clear, and intuitive interpretation of entropy.

In addition to the advantages of the definition of entropy based on SMI, I will present in Chapter 1 a new formulation of the Second Law based on *probability* rather than on *entropy*. The traditional formulation of the Second Law in terms of entropy is valid for *isolated systems*. On the other hand, the formulation of the Second Law in terms of probability is valid for any well-defined thermodynamic system: isolated (constant E, V, N), isothermal-isobaric (constant T, P, N), or any other thermodynamic system. This formulation states that when a constraint is removed from any well-defined system at equilibrium, the system will evolve into a new equilibrium state having an overwhelmingly higher probability.

For an isolated system, the entropy *increases* as a result of such a process. For an isothermal-isochoric system, the Helmholtz energy decreases, and for an isothermal-isobaric system, the Gibbs energy decreases. Thus, the traditional formulation of the Second Law based on entropy is replaced by the more general formulation based on probability. More on this may be found in Chapter 1.

The second chapter surveys the most common (and some not-so-common) interpretations (or descriptors, or metaphors) of entropy. In each case at least one example is presented to debunk the "claimed" interpretation of entropy. What is left is a single, valid, and proven interpretation of entropy as well as of the Second Law.

In the third chapter, I chose a few examples which I am familiar with, and which I believe represent the usefulness of both entropy and the Second Law.

Toward the end of Chapter 3, I will mention very briefly some common misuses of entropy and the Second Law. Recently, the market has been flooded with popular science books which harp on entropy's and the Second Law's connection with life, the universe, black holes, and more. It seems to me that many authors feel that they are at liberty to write whatever comes to their minds, and whether consciously or unconsciously, they succumb to the temptation of writing something knowing that they will be "immune" from being proven wrong. This "immunity" from criticism probably stems from von Neumann's statement that "no one knows what entropy really is."

This book is addressed to anyone who is curious about what is considered by many scientists to be the "most mysterious" law in physics. It can be read by the uninitiated, or by practically anyone who is a novice or who does not have any knowledge in mathematics or physics.

It is also addressed to students of thermodynamics who were exposed to entropy, but were left baffled as to its meaning. Finally, it is also addressed to teachers of thermodynamics, authors of books on thermodynamics, and researchers who apply thermodynamics. To all of them, I can offer a new approach based on a solid *definition* of entropy, which leads to a sound *interpretation* of entropy, and which is implemented in interpreting some simple processes.

To all the readers of this book I can promise a crystal clear definition of entropy and an understanding of the Second Law based on plain common sense, leaving no traces of mystery which has enshrouded entropy and the Second Law for over a hundred years.

The style of this book is largely descriptive and non-mathematical. Some technical details are relegated to the notes, while some mathematical equations which are of fundamental importance are shown in boxes.

Arieh Ben Naim

Department of Physical Chemistry
The Hebrew University of Jerusalem
Jerusalem, Israel
Email: ariehbennaim@gmail.com
URL: ariehbennaim.com

Acknowledgments

I am grateful to John Anderson, David Avnir, Zeev Aizenstat, Diego Cassadei, Claude Dufour, Robert Engel, David Gmach, Ori Gidron, Jose Angel Sordo Gonzalo, Robert Hanlon, Richard Henchman, Oded Kafri, Zvi Kirson, Jonathan Langford, Bernard Lavenda, Azriel Levy, Kent Mayhew, Mike Rainbolt, Eric Szabo, David Thomas, and Harry Xenias for reading parts or the entire manuscript and offering useful comments.

As always, I am very grateful for the gracious help I got from my wife, Ruby, and for her unwavering involvement in every stage of the writing, typing, editing, re-editing, and polishing of the book.

Test Yourself Before Reading This Book

I believe that in order to understand *entropy* it is absolutely essential to understand the Shannon's measure of information (SMI), and to make a clear-cut distinction between entropy and SMI. In order to understand SMI, it is absolutely essential to understand the difference between *information* and *measure* of *information*.

Of course, not everyone must understand what entropy is. However, I believe that any educated person, scientist as well as non-scientist, should be familiar with the basic notion of SMI. Once you get familiar with SMI, you will effortlessly understand what entropy is.

Before you start immersing yourself in this book, I suggest that you do the following test or exercise. Usually, one does an exercise in order to assess one's understanding of a newly learned subject. Here, I urge you to do an exercise *before* and *after* reading the book. By doing so, you will be able to compare your answers, and assess the extent of your understanding of the three main concepts discussed in this book: information, SMI, and entropy.

The Qualitative Test

Consider the following items of information:

1. It rained in Jerusalem yesterday.
2. It is raining in Jerusalem today.
3. It will rain in Jerusalem tomorrow.
4. It will be raining next week.
5. It will be raining all of next week, from Sunday to Saturday.
6. It will be raining all the days next week, starting on Sunday, then also on Monday, also on Tuesday, and all the days of Wednesday, Thursday, Friday, and Saturday.
7. No one was injured in a car accident.
8. Two people were injured in a car accident.
9. Twenty five people were injured in a car accident.
10. 25 people were injured in a car accident.

Read carefully *all* the ten items of information before you start answering the following questions:

(i) What is the information you got?
(ii) How much new information did you get?
(iii) How surprised were you when you got this information?
(iv) What is the *size* of the information you got?

I urge you to write down your answers, before and after reading this book. Only qualitative answers are required.

The Quantitative Tests

Consider the following items of information:

Case A

1. I tell you that I threw a die, and that its probability distribution is: $P_1 = P_2 = P_3 = P_4 = P_5 = P_6 = \frac{1}{6}$ (i.e. all of the possible outcomes $1, 2, \ldots, 6$ are equally likely to occur)

2. The outcome was "4"
3. The outcome was "1"
4. The outcome was "even"

Case B

1. I tell you that I threw a die, and that its probability distribution is:
   $P_1 = \frac{1}{2}, P_2 = 0, P_3 = 0, P_4 = \frac{1}{4}, P_5 = \frac{1}{4}, P_6 = 0$
2. The outcome was "4"
3. The outcome was "1"
4. The outcome was "even"

Case C

1. I tell you that I threw a die, and that its probability distribution is:
   $P_1 = 1, P_2 = P_3 = P_4 = P_5 = P_6 = 0$
2. The outcome was "4"
3. The outcome was "1"
4. The outcome was "even"

Case D

1. I tell you that I threw a die, and I do not know the distribution: $(P_1, \ldots, P_6)$
2. The outcome was "4"
3. The outcome was "1"
4. The outcome was "even"

Read carefully *all* the items of information above, then answer the following questions:

(i) What is the information you got?
(ii) How much uncertainty is involved in the outcomes of the die?
(iii) How surprised were you when you got this information?
(iv) Can you estimate the Shannon measure of information for each of the items in the test?

I urge you to write down your answers, before reading this book. Although this is a "quantitative" text, only qualitative answers are required.

Two More Qualitative Tests

Table I

If you know something about entropy, examine the truthfulness of the following statements.

Statement	True	False
Entropy tends to increase.		
Entropy of the universe always increases.		
Entropy increases in a spontaneous process.		
Entropy of a well-defined system tends to increase.		
Entropy of a well-defined isolated system tends to increase.		
Mixing is viewed as disordering — therefore mixing increases entropy.		

Table II

If you know something about information theory, examine the truthfulness of the following statements.

Statement	True	False
Information always increases.		
Information in the universe is constant.		
Information increases in a spontaneous process.		
Information of a well-defined system is constant.		
Information in an isolated system always increases.		

1

The Various Definitions of Entropy and the Second Law of Thermodynamics

1.1 Introduction

This chapter introduces three definitions of entropy. The first is referred to as either the thermodynamic, experimental, macroscopic, or non-atomistic definition. This definition originated in the 19th century, stemming from the interest in heat engines. The introduction of entropy into the vocabulary of physics is due to Clausius. In reality, Clausius did not *define* entropy itself, but only changes in entropy. Clausius' definition, together with the Third Law of Thermodynamics, led to the calculation of "absolute values" of the entropy of many substances.

The second definition is attributed to Boltzmann. This definition is sometimes referred to as either the microscopic definition or the atomistic definition of entropy. It relates the entropy of a system to the number of accessible microstates of a thermodynamic system characterized macroscopically by the total energy E, volume V, and total number of particles N (for a multi-component system N may be reinterpreted as the vector $(N_1, \ldots, N_c)$, where N_i is the number of atoms of type i.) The extension of Boltzmann's definition to systems characterized by independent variables other than (E, V, N) is due to Gibbs. Gibbs also introduced the idea of an ensemble of systems and showed how one can reformulate the Second Law of Thermodynamics for systems characterized by other sets of independent variables such as T, V, N or T, P, N. We shall briefly discuss the idea of ensembles in Section 1.6. For each of these systems one can define a *potential function* (or a fundamental function) which attains an extremum value at equilibrium. For instance, the Helmholtz energy attains a minimal value for a system characterized by the variables T, V, N. Similarly, the Gibbs energy attains a minimal value for a system characterized by the variables T, P, N. Ultimately, all these reformulations of the Second Law can be traced back to the formulation in terms of maximal value of the entropy of an isolated system (i.e. E, V, N constants).

The Boltzmann definition seems to be completely unrelated to the Clausius definition, yet it is found that for all processes for which entropy changes can be calculated by using Boltzmann's definition, the results agree with entropy changes, which are calculated using Clausius' definition. Although there is no formal proof that Boltzmann's entropy is equal to the thermodynamic entropy, it is widely believed that this is true.

The third definition is based on the Shannon's measure of information (SMI). It may also be referred to as the microscopic or the atomistic definition of entropy. However, this definition of entropy, as well as the Second Law, is very different from Boltzmann's definition. This is also the only definition which provides a simple, intuitive, and meaningful interpretation of entropy and the Second Law.

We also note that calculations of entropy changes based on the SMI definition agree with those calculations based on Clausius' as well as on Boltzmann's definition. Unlike Boltzmann's definition, the SMI definition does not rely on calculations of the number of accessible states of the system. It provides directly the entropy function of an ideal gas, and by extension, also the entropy function for a system of interacting particles. Other advantages of this definition are discussed at the end of Section 1.5.

In the following sections we shall present the different definitions of entropy, and the different formulations of the Second Law in a qualitative, non-mathematical language. In some cases, we present some mathematical details in the notes to each chapter. However, for a full mathematical exposition of the subject we shall refer to the appropriate literature.

After surveying the three definitions of entropy and the various formulations of the Second Law of Thermodynamics, we also briefly mention a few misdefinitions of entropy and misformulations of the Second Law of Thermodyamics. Sometimes, an author might make statements such as: "The entropy is a measure of...," or "the Second Law is the tendency of a system to proceed toward...." In such cases it is difficult to tell whether the author intends to either define or interpret entropy. In this chapter, we shall deal only with *definitions* of entropy. We shall further examine critically some common *interpretations* of entropy and the Second Law in Chapter 2.

1.2 The Clausius Definition and Its Extensions

In this section, we start with the early considerations regarding heat engines which led Clausius to introduce the concept of entropy and formulate the Second Law of Thermodynamics. We then briefly discuss the formulation of the Second Law in other systems of interest in thermodynamics.

1.2.1 *Heat Engines and Carnot's Efficiency*

Traditionally, the birth of the Second Law is associated with the name Sadi Carnot (1796–1832). Although Carnot himself did not formulate the Second Law, his work laid the foundations on which the Second Law was formulated a few years later by Clausius and Kelvin. Carnot was interested in heat engines, more specifically, in the *efficiency* of heat engines.

Heat engines were supposed to do the *work*, which people used to do with their bare hands and raw muscles. Basically, you can think of "heat engines" as an analog of waterfall engines. In waterfalls, water cascades from higher to lower levels. On its way down, the water can rotate a turbine. We can use this rotation to our advantage like plowing a field or generating electricity. Likewise, *heat* flows spontaneously from a high temperature to a low temperature. On its way down, this flow of heat can be exploited by us for doing some work, like running a train or lifting weights from low to high levels.

In the 19th century, scientists believed that heat is a kind of fluid called *caloric* that flows from a higher to a lower temperature. Today, this "caloric theory" is considered obsolete. We shall still use the term "heat flow" to mean heat transferred.

Qualitatively, think again of a waterfall, and imagine that from some quantity of water falling from h_2 to h_1, you can do some useful work. Likewise, for a given amount of heat "falling" from the higher temperature T_2 to a lower temperature T_1,[1] you can do some useful work (Figure 1.1).

Note, however, that "falling of water" and the "flow of heat" are governed by very different laws. The first is governed by Newton's Gravitation Law, the second, by the Second Law of Thermodynamics (see also Note 1 to the preface).

Carnot was interested in the *efficiency* of a heat engine, how much useful work one can get from a given amount of heat that flows from the higher temperature T_2 to the lower temperature T_1. Carnot found, somewhat unexpectedly, that there is a limit on the efficiency of a heat engine operating between two temperatures T_2 and T_1.[1] This finding was not a formulation of the Second Law, but it sowed the seeds for the inception of the Second Law.

The Various Definitions of Entropy | 9

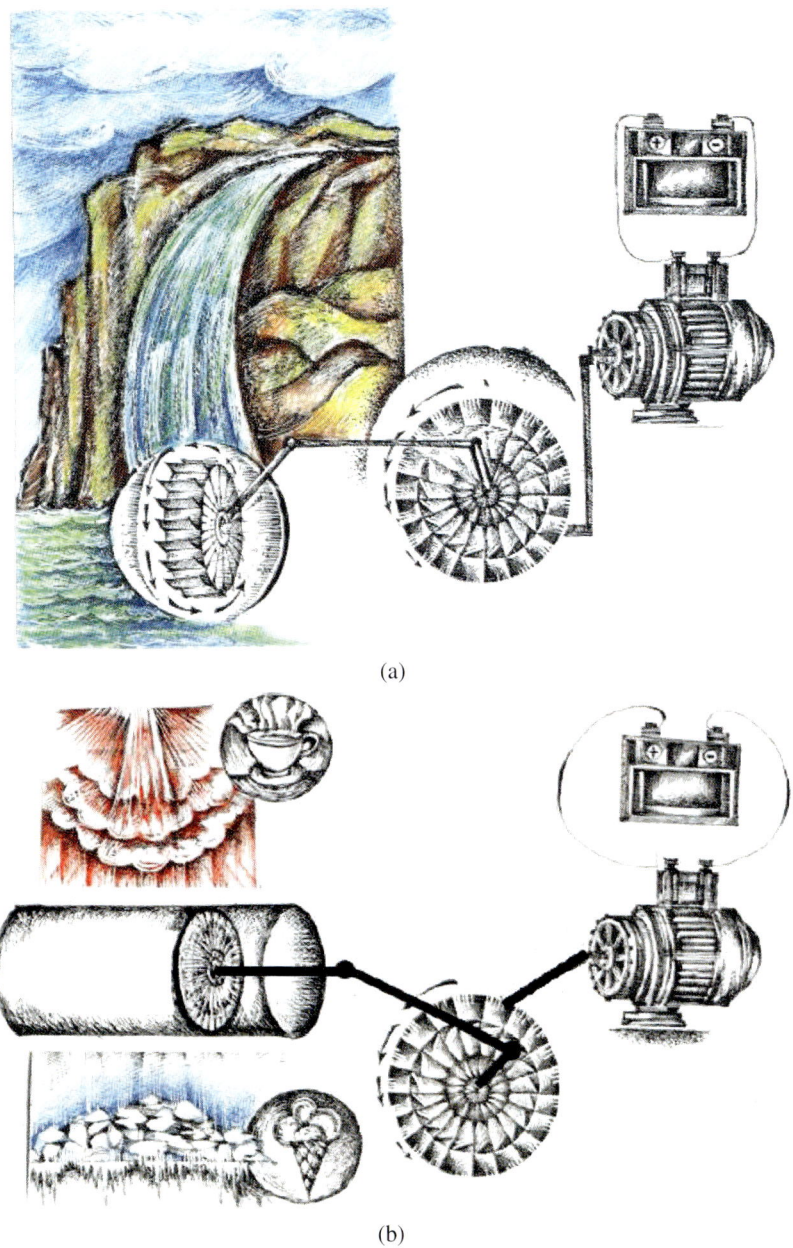

(a)

(b)

Fig. 1.1. (a) An artist rendition of (a) "Water fall," and (b) "Heat fall."

The seeds sowed by Carnot sprouted in different directions which led to the different formulations of the Second Law.

1.2.2 *Clausius' Definition of Entropy*

Basically, Clausius observed, as every one of us does, that there are many processes that occur in nature spontaneously and always in one direction. Examples abound.

- Take gas in a small box, and open it within a larger empty box. You will *always* observe that the gas will expand and fill the entire volume of the bigger box (Figure 1.2a).
- Take two gases, say argon and neon, in different compartments separated by a partition. Remove the partition, and you will always observe a spontaneous mixing of the two gases (Figure 1.2b).
- Take two identical pieces of iron, one at 300°C and the second at 100°C. Bring them to thermal contact. You will always

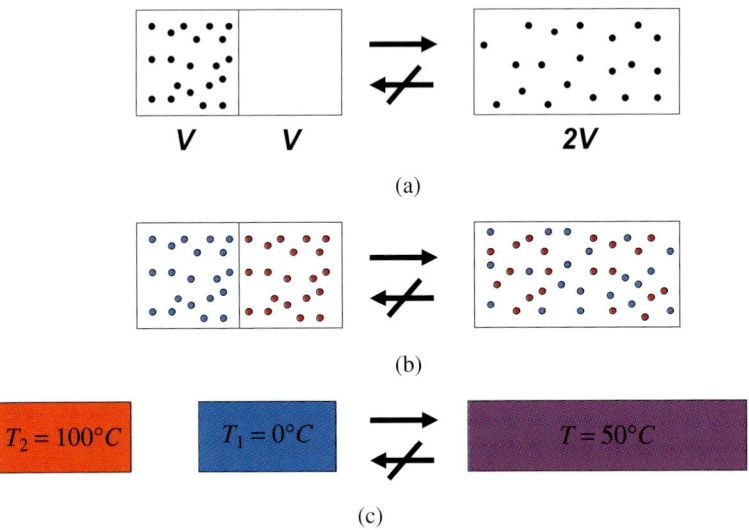

Fig. 1.2. Three spontaneous processes in isolated systems for which the entropy increases: (a) expansion of an ideal gas, (b) mixing of two ideal gases, and (c) heat transfer from a cold to a hot body.

observe a spontaneous flow of heat from the body with the higher temperature to the body with the lower temperature. The hotter body will cool down; the colder body will heat up. At equilibrium the two bodies will have a uniform temperature of 200°C (Figure 1.2c).

In all of these examples we *never* observe the reverse process spontaneously; the gas *never* condenses into a smaller region in space, the two gases *never* un-mix spontaneously after being mixed,[2] and heat *never* flows from the cold to the hot body.[3] Note carefully that I italicized the word *never* in the previous sentences. Indeed, we *never* observe any of these processes occurring spontaneously in the reverse direction. For this reason, the processes shown in Figure 1.2 (as well as many others) are said to be *irreversible*. However, one should be careful with the use of the words "reversible" and "irreversible" in connection with the Second Law. There are several, very different, definitions assigned to these words. For more details, see Ben-Naim (2011a, 2015a, 2016a). Here, we point out two possible meanings of the term "irreversible."

1. We *never* observe the final state of any of the processes in Figure 1.2 returning back to the initial state (on the left hand side of Figure 1.2) spontaneously.
2. We *never* observe the final state of any of the processes in Figure 1.2 returning back to the initial state, and *staying* in that state.

In case 1, the word *never* is used in "practice." The system can go from the final to the initial state. In this case, we can say that the initial state will be *visited*. However, such a reversal of the process would occur once in many ages of the universe. Therefore, this is *practically* an irreversible process; we will "never" observe such a reversal in practice.

In case 2, the word *never* is used in an absolute sense. The system will never go to its initial state, and stay there!

In Section 1.5 we discuss a few processes of expansion with small numbers of particles. In such systems when we remove the partition between the two chambers, the particles will move to occupy the larger volume spontaneously. However, from time to time we will observe all the particles in one chamber (either the left, or the right one). However, once they are in one chamber (and this will occur with decreasing probability as the number of particles increase), the particles will expand again to occupy all the available volume. In order to *stay* in one chamber, the partition should be replaced at its original position. Of course, the particles staying in the one chamber will never (in an absolute sense) occur spontaneously.

The distinction between these two "irreversibilities" will be important in connection with the formulation of the Second Law. Going back to the three processes shown in Figure 1.2, we can ask *why* these processes always occur in one direction. Is there a law of nature that dictates the direction of the unfolding of these processes? Look again at the three processes depicted in Figure 1.2. Take note that these are quite *different* processes and that it is far from being clear that they are all governed by the same law. Perhaps, there is one law for the spontaneous expansion of a gas, another for the spontaneous mixing of two gases, and still another for the spontaneous flow of heat from the hot to the cold body.

It was Clausius who realized the common principle underlying all these processes, and postulated that there is only one law that governs all these processes. Even before formulating the Second Law, Clausius' postulate was an outstanding achievement considering the fact that none of these processes was understood. You watched how a colored gas expands and fills a larger volume. You watch a drop of blue ink mixing with a glass of water coloring the entire liquid. You watch the hot body cooling and the cold body heating. You watch all of these with your *macroscopic eyes*, but you have no idea what *drives* these processes, what goes on *inside* the system you are watching. Such an insight was not even possible before the atomic nature of matter was embraced by the scientific community which allowed us to use our "*microscopic eyes*" to "see" what is going on when such processes occur. "Seeing," even with our microscopic eyes, is one thing, and *understanding* what we see is quite another.

Clausius started from one particular process: the spontaneous flow of heat from a hot to a cold body. Based on this specific process, Clausius defined a new quantity which he called *entropy*. Let $dQ > 0$ be a *small quantity* of heat flowing into a system, being at a given temperature T. The *change* in *entropy* was defined as[4]

$$dS = \frac{dQ}{T}$$

The letter d stands here for a very small *quantity*, and T is the absolute temperature. Q has the units of *energy*, and T has the units of *temperature*. Therefore, the entropy change has the units of *energy* divided by units of *temperature*. The quanity of heat dQ *must* be very small, such that when it is transferred into, or out, from the system, the temperature T does not change. If dQ is a finite quantity of heat, and you transfer it to a system which is initially at a given T, the temperature of the system might change, and therefore the change in entropy will depend on both the initial and the final temperature of the system. Note carefully that this equation does not define *entropy* but only *changes in entropy* for a particular process, i.e. a *small* exchange of heat ($dQ > 0$ means heat flows into the system, $dQ < 0$ means heat flows out of the system). There are many processes, a few of which are shown above in Figure 1.2, which do not involve heat transfer (in Figure 1.2c we consider the two parts of the system as one isolated system; within this system there is a flow of heat from one part to the other). Yet, from Clausius' definition, and the postulate that the entropy is a *state function*, one could devise a *path* leading from one state to another, for which the entropy change can be calculated.

It is not uncommon to refer to the equation $dS = dQ/T$ as Clausius' *definition* of entropy.[4] In fact, this equation does not define entropy, nor *changes* in entropy for a general process (e.g. expansion of an ideal gas). One must also add to this definition that the entropy is a state function. See Section 1.2.3 below.

Initially, Clausius formulated a "restricted" Second Law, namely that heat does not flow spontaneously from a cold body to a hot body. However, later he postulated that there exists a quantity he called entropy, which is assigned to any macroscopic system, and when a spontaneous process occurs, the entropy always increases. This was the birth of the Second Law of Thermodynamics. This law introduced a new quantity to the vocabulary of physics, and at the same time brought together many processes under the same umbrella.

The extraordinary achievement of Clausius was the enormous generalization from a few spontaneous processes to *any* spontaneous process. It should be stressed here that the formulation of the Second Law in terms of entropy is valid only for isolated systems, i.e. systems having a constant energy, volume, and number of particles. For other systems, say constant P, T, N, the entropy can go up or down. We shall discuss these formulations of the Second Law in Section 1.5.

Soon after Clausius formulated the Second Law, scientists proved that various particular formulations were all equivalent to each other. The proofs of the equivalence appear in any textbook of thermodynamics. Look again at the three processes depicted in Figure 1.2. These are very different processes, but they are governed by the same law, the Second Law of Thermodynamics. Today, we can calculate the change in entropy and we find that, whenever a spontaneous process occurs in an *isolated* system, the entropy of the system always increases. Before we continue, we must emphasize that by "entropy changes" we mean differences in entropy between two equilibrium states. We do not know how to calculate the entropy change for any spontaneous process.

At this point, we pause to discuss the important concept of equilibrium state. Experimentally, we know that systems consisting of a huge number of particles can be described by a few parameters. For instance, a gas consisting of N atoms of argon can be described by its pressure and its temperature. This description is referred to as the *thermodynamic* or *macroscopic* state of the system. Clearly, a macroscopic state does not specify the microscopic states of the system. For these, we need to know the locations and velocities of a huge number of particles, $N \approx 10^{23}$.

We also know that there exist states for which the thermodynamic parameters, say temperature, pressure, or density, do not change with time. These states are called equilibrium states.

It should be stressed, however, that there is no general definition of equilibrium which applies for all systems. Callen (1985) introduced the existence of the equilibrium states as a postulate. He also emphasized that any definition of an equilibrium state is necessarily circular.

In practice, we find many systems, like glasses, for which the parameters describing the system seem to be unchanged with time. Yet, they are not equilibrium states. But for all our purposes, we assume that every well-defined system, say having a fixed energy E, volume V, and number of particles N, will tend to an equilibrium state. At this state, the entropy of the system is defined as well as other thermodynamic quantities.

We mentioned above the entropy for an isolated system (i.e. fixed E, V, and N). There exists no strict isolated system. However, such systems are convenient for the construction of thermodynamics, as well as for statistical mechanics.

As for the choice of the term "entropy," Clausius explained:[5]

> I propose, accordingly, to call S the **entropy** of a body, after the Greek word **'transformation.'** I have designedly coined the word entropy to be similar to **energy**, for these two quantities are so analogous in their physical significance, that an analogy of denominations seems to me helpful.

The choice of this term was not entirely appropriate.[6] We shall further discuss this issue in Section 2.2.1. However, during the time it was chosen, the meaning of entropy was not clear. It was a well-defined quantity, and one could calculate changes in entropy for many processes without giving a second thought to the *meaning* of entropy. Perhaps, there is no "deeper" meaning to entropy; perhaps entropy is just another physical quantity, such as volume and energy which do not have any "deeper" meaning. In fact, there are many scientists who use the concept of entropy successfully and who do not care for the meaning of entropy, or if it has a meaning at all. As Battino has stated:[7]

> ... all of the useful relationships in thermodynamics can be developed and used without any knowledge of the existence of atoms and molecules.

At this stage we might be content to have a quantity that is well defined in thermodynamics. The term itself might not be appropriate,[6] but it has been with us for over a hundred years, and whatever its meaning is, in either ancient or modern Greek, it will probably stay with us. However, once we do this we should be careful not to use the same term for another concept as this will cause a huge confusion. That is exactly what happened when Shannon himself acceded to the suggestion to rename his measure entropy. We shall further discuss this aspect of entropy in Section 1.4.

Notwithstanding the enormous success and the generality of the Second Law, Clausius made one further generalization of the Second Law:

The entropy of the universe always increases.

This formulation can be said to be an unwarranted *overgeneralization*. We shall further discuss the fallacy of this overgeneralization in Section 3.12.

It should be noted that in each formulation of the Second Law there is some degree of generalization which cannot be verified. Here we mentioned a generalization for the whole universe for which the entropy cannot be defined.

1.2.3 *Entropy Change in an Expansion of an Ideal Gas*

Following Clausius' definition, it was soon realized that although thermodynamics did not provide an absolute value for the entropy, it was recognized that the entropy is a *state function*. This means that the entropy is *defined* for any well-defined system at *equilibrium*. From this realization, it follows that if a system changes from one well-defined state to another well-defined state, the entropy change in this process is fixed by the initial, and the final, states of the system. It is here that the concept of reversible and irreversible processes entered into thermodynamics.[8] Instead of discussing these concepts in abstract terms, we provide a simple example: the expansion of an ideal gas from a volume V to a volume $2V$ (see Figure 1.2a).

Consider the process depicted in Figure 1.2a. We start with a system of N simple particles, say argon atoms in a volume V, and having total energy E. We assume that the gas is *ideal*, i.e. we neglect intermolecular interactions. Such a system obeys the equation of state $PV = Nk_BT$, where P is the pressure, T the temperature, and k_B the Boltzmann constant. We also assume for simplicity that the total energy of the system is the sum of the kinetic energies of all its particles.

We remove the partition between the two compartments in the system, and we observe the expansion of the gas to occupy the entire, new volume $2V$.

How do we calculate the change in entropy for this process? Recall that Clausius' definition is $dS = dQ/T$. In the process depicted in Figure 1.2a, there is no transfer of heat into, or out, from the system. One might conclude that since $dQ = 0$, then also $dS = 0$, i.e. no change in entropy. This conclusion is, however, wrong. The equation $dS = dQ/T$ applies for a *specific* process of transferring of a small amount of heat to a system at constant temperature, not to any process.

Note carefully that for the specific process of transferring a small amount of heat, dQ, it is appropriate to say that the associated entropy $dS = dQ/T$ *flows* into, or out, from the system. This terminology is not advisable, however. Not every change of entropy is due to "flow," and certainly one cannot say that entropy itself is a quantity that "flows." In fact, even for a small amount of transferring of heat, the distribution of energy in the system is changed; we can say that as a result of the change in the distribution, entropy increases or decreases, but it does not "flow." As we shall see soon, the Boltzmann entropy cannot be said to be flowing into, or out, from the system; the number of configurations or microstates can increase or decrease, but it does not "flow." This will be *a fortiori* true for the entropy definition based on the Shannon's measure of information. For more on that see Ben-Naim (2015a).

To calculate the (finite) change in entropy between two well-defined states, we must use the idea that the entropy is a *state function*. Here we show an example of how one can calculate the finite change in entropy using the Clausius definition, plus the idea that the entropy is a state function.

The process depicted in Figure 1.2a may be described in the pressure-volume (PV) diagram shown in Figure 1.3. The initial state A is P_1, V_1, T (here $V_1 = V$ and $P_1 = Nk_BT/V_1$), and the final state is P_2, V_2, T (where $V_2 = 2V$, and $P_2 = Nk_BT/V_2 = P_1/2$). Note that the temperature does not change in this process. This is true for an ideal gas. Because the gas is ideal, the doubling of the volume of the system causes a reduction of the pressure from P_1 to $P_1/2$.

Thus, on the PV diagram we have two points: the initial P_1, V_1 and the final states P_2, V_2. When we remove the partition between the two compartments, the system will move from the initial state to the final state. We cannot describe in this diagram any intermediate point, simply because during the process the system does not proceed through well-defined points.

In order to calculate the change in entropy in this process, we exploit the fact that entropy is a *state function*. This means that if we know the initial and the final *states* of the system, the change in entropy is determined. In our case, we write $\Delta S = S(P_2, V_2, T) - S(P_1, V_1, T)$. This difference is independent of the *path* along which the system moves from the initial state to the final state.

Thus, we know that ΔS for the process in Figure 1.2a is well defined. However, we still have to find a way to calculate ΔS for this process. Thermodynamics offers a truly ingenious way of doing this calculation. Devise *any* process leading from the initial to the final process for which we *know* how to calculate ΔS, and there we are! ΔS will be the same for whatever process leads from the point $P_1 V_1$ to the point $P_2 V_2$.

Now, we use the Clausius definition ($dS = dQ/T$) along with the first law of thermodynamics which we write as $dE = dQ + dW$, where dE is the change in the internal energy of the system, dQ is the heat flow into the system, and dW is the work done on the system. Note again that the internal energy E, which in our example is the sum of the kinetic energies of all the particles, is a *state function*. Hence, in going from one well-defined state to another, the change in energy ΔE is determined. On the other hand, dQ and dW stand for *small* quantities of heat and work. Neither heat nor work is a state function. This means that work can be done on, or by, the system; heat can be transferred to, or from, the system. Both heat and work are forms of energies which can be *transferred* to, or from, the system.

In the process of expansion of an ideal gas (see Figure 1.2a), the internal energy does not change, i.e. $\Delta E = 0$. From Clausius' definition and the first law we get, for any small change in the system,

$$dE = dQ + dW = 0 = TdS - pdV$$

Here, dW is the work done by the system while expanding. This equation tells us that if we find a path along which the system moves from the initial to the final state, and for which we can calculate the *work* done by the system, or on the system, then we can calculate the change in the entropy of the system. It is here that the concept of *reversible* process is introduced. We prefer to use the concept of a *quasi-static* process instead. The details of the calculations are given in Note 9. The result is

$$\Delta S(expansion) = Nk_B \ln 2$$

This is a remarkable result. The process depicted in Figure 1.2a does not involve any transfer of heat into, or from, the system, yet we could use the Clausius' definition along with the fact that the entropy is a state function to calculate the finite entropy change (ΔS) for this process. We did this by constructing a *path* consisting of a very dense sequence of equilibrium states. This is referred to as a quasi-static process[8] (see Figure 1.3).

It should be stressed that by "reversible" in this process, we mean that the process can be *reversed* along the *same* sequence of states along the way from A to B. At each point, *including* the initial and the final points, the system is assumed to be at equilibrium. It is only for equilibrium states that the entropy is defined. This is the meaning of the entropy being a *state function*.

Thermodynamics does not tell us how exactly the entropy of a system depends on the parameters that characterize the system, such as temperature, pressure, volume, etc. Yet we can deduce that for an isolated system characterized by a fixed energy E, volume V, and number of particles N (assuming for simplicity that the system contains only one kind of particles), the function $S = S(E, V, N)$ must be a monotonically increasing function of E, V, and N. Furthermore, from the Second Law it follows that this function must be concave downwards, i.e. the slope of S as a function of E, V, or N must be *positive* and *decreasing* as E, V, or N increases. We shall see this behavior when we obtain the entropy function of ideal gas.

We can summarize this section as follows. Thermodynamics provides a method for calculating the entropy change between any two *equilibrium states*. It does not tell us the *value* of the entropy of a system at equilibrium. It only states that there exists such a *state*

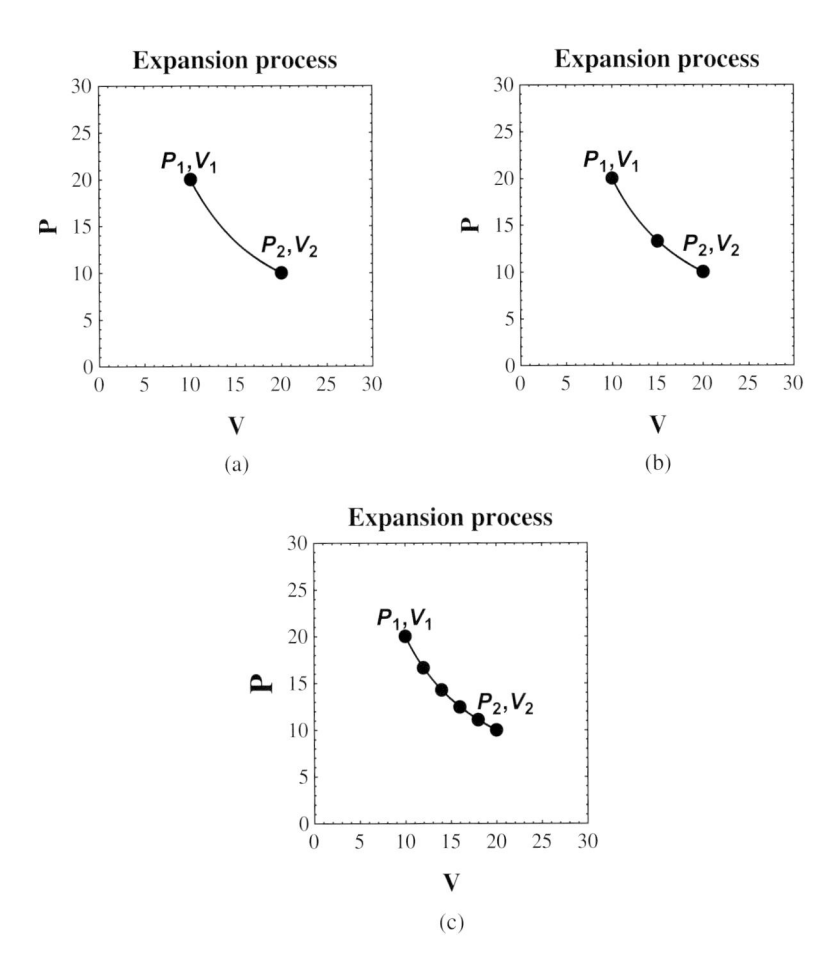

Fig. 1.3. Pressure-volume (PV) diagram: (a) only the initial and the final states shown, (b) one intermediate state added, and (c) many intermediate states added.

function. When the state of the system is defined by the variables E, V, and N, the entropy function must be a monotonically increasing function of each of these variables, and in addition this function must have a negative *curvature* (or concave downwards) with respect to these variables. Furthermore, the Second Law states that in any process occurring spontaneously, in an isolated system in which the system moves from one equilibrium state to a new equilibrium state, the entropy must increase. Thermodynamics does not reveal to us *what* the meaning of entropy is, nor *why* it always changes in one direction only.

1.3 The Microscopic Definition of Entropy

1.3.1 *Introduction*

Toward the end of the 19th century, the atomistic theory of matter was firmly consolidated. The majority of scientists believed — yes, it was still a belief — that matter consists of small units called atoms and molecules. A few persistently rejected that idea arguing that there is no proof of the existence of atoms and molecules; no one has seen any atom or a molecule! Therefore, they justifiably claimed that the existence of atoms and molecules was a mere speculation.

On the other hand, the so-called *kinetic theory of heat*, which was based on the assumption of the existence of atoms and molecules, had scored a few impressive gains. First, the pressure of a gas was successfully explained as arising from the molecules bombarding the walls of the container. Then came the interpretation of temperature in terms of the kinetic energy of the molecules. That was a remarkable achievement that supported and lent additional evidence for the atomic constituency of matter, but fell short of a proof.

Remember that both pressure and temperature are measurable quantities. We can feel both of them on the tip of our fingers. Neither the measurements nor our feelings give us any hint that these quantities are manifestations of the motions of huge numbers of tiny particles.

Furthermore, the concept of *heat*, which was believed to be a kind of fluid that flows from a hot to a cold body, was also interpreted in terms of the energies of all the individual molecules. Under this interpretation, the First Law of Thermodynamics is nothing but an extension of the principle of conservation of energy, which now also embraces heat, or thermal energy, as just another form of energy.

Thus, while the kinetic theory of heat was successful in explaining the concepts of pressure, temperature, and heat, it was left lagging behind entropy and the Second Law of Thermodynamics.

1.3.2 *Boltzmann's Definition of Entropy*

Boltzmann picked up the challenge and defined entropy in terms of the *total number of microstates* of a system consisting of a huge number of particles, but characterized by the macroscopic parameters of energy E, volume V, and number of particles N.

What are these "number of microstates," and how are they related to entropy?

Consider a gas consisting of N simple particles in a volume V; each particle may be at some location R_i and have some velocity v_i. By simple particles we mean particles having no internal degrees of freedom. Atoms of elements such as argon, neon, and the like are considered as simple. They all have internal degrees of freedom, but these are assumed to be unchanged in all the processes we discuss here. Assuming that the gas is very dilute so that interactions between the particles can be neglected, then, all the energy of the system is simply the sum of the kinetic energies of all the particles.

Imagine that you could have microscopic eyes, and you could see the particles rushing incessantly, colliding with each other, and with the walls from time to time. You will be impressed that there are infinite *configurations* or *arrangements* of the particles which are consistent with the requirements that the total energy is constant, and that they are all contained within the box of volume V. One such configuration is shown in Figure 1.4. Each particle is specified by its location R_i and its velocity v_i.

Without getting bogged down with the question of how to define and how to calculate the total number of arrangements, it is clear that this is a huge number, far "huger" than the number of particles which is of the order $N \approx 10^{23}$. Boltzmann postulated the relationship which

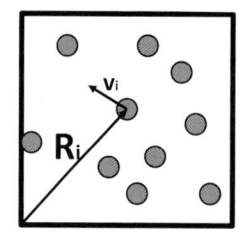

Fig. 1.4. Particles in a box of volume V. Each particle has a locational and a velocity vector.

is now known as the Boltzmann entropy

$$S = k_B \log W$$

where k_B is a constant, now known as the Boltzmann constant, and W is the number of microstates of the system. Here, log is the natural logarithm. At first glance, Boltzmann's entropy seems to be completely different from Clausius' entropy. Nevertheless, in all cases for which one can calculate changes of entropy one obtains agreements between the values calculated by the two methods.

1.3.3 *Objections to Boltzmann's Entropy*

Boltzmann's entropy was not easy to swallow, not only by those who did not accept the atomic theory of matter, but also by those who accepted it. The criticism was not focused so much on the definition of entropy, but more on the formulation of the Second Law of Thermodynamics. Boltzmann explained the Second Law as a probabilistic law. In Boltzmann's words:[10] "...a system...when left to itself, it rapidly proceeds to disordered, most probable state."

This statement, especially the phrase "most probable state," was initially shocking to many physicists. Probability was totally foreign to physical reasoning. Physics was built upon deterministic and absolute laws; there were no provisions for exceptions. The macroscopic formulation of the Second Law was absolute — no one has ever observed a single violation of the Second Law. Boltzmann, on the other hand, insisted that the Second Law is only *statistical*; entropy increases *most* of the time, not *all* the time. The decrease in entropy is not an *impossibility* but is only highly improbable.[10]

At the time when Boltzmann proclaimed the probabilistic approach to the Second Law, it seemed as if this law was somewhat *weaker* than the other laws of physics. All physical laws are absolute and no exceptions are allowed. The Second Law, as formulated by Clausius, was also absolute. On the other hand, Boltzmann's formulation was not absolute — exceptions were allowed. It was much later realized, however, that the admitted non-absoluteness of Boltzmann's formulations of the Second Law, is in fact more *absolute*, than the absoluteness of the macroscopic formulation of the Second Law, as well as of any other law of physics for that matter.[11]

There were essentially two "paradoxes," the so-called "reversibility paradox" and the "recurrence paradox."[12]

Basically, the paradoxes do not stem from Boltzmann's *definition* of entropy as stated above, but from a function Boltzmann defined (and showed that it always decreases with time). This function was known as the H-function and the theorem Boltzmann proves was referred to as Boltzmann's H-theorem.

As we shall see, the H-function is a particular case of the Shannon's measure of information.

Indeed, it is true that H decreases with time until it reaches a minimum at equilibrium. The function $-H(t)$ looks like entropy but it is not the entropy of the system.[12]

1.3.4 *Absolute Values of Entropy and the Third Law*

As we noted in Section 1.2, Clausius did not define the entropy function, nor did he provide a method of calculating the *value* of the entropy for any system at equilibrium. This could be achieved by using the Third Law of Thermodynamics. The application of the Third Law to calculate the absolute values of the entropy from experimental data (on heat capacity and heat of phase transitions) will be discussed in

Section 3.2. This procedure involves integration of experimental data from temperatures as close as possible to absolute zero, to the final state of interest. This is referred to as the experimental absolute value of the entropy.

On the other hand, if we assume that the Boltzmann definition is valid at all temperatures, including absolute zero ($T = 0K$), then Boltzmann's equation implies that at absolute zero temperature, the number of states of the system is one, i.e. the entropy of the system is zero.

Indeed, for many systems for which one calculates the entropy, both from Boltzmann's equation and from experimental data, one gets good agreement between the two. However, there are many exceptions, and we shall discuss some examples which are well understood in Section 3.2.

1.3.5 Extension of the Second Law to Non-Isolated Systems

The entropy formulations of the Second Law states that if we remove a constraint in a *constrained equilibrium state* of an isolated system, the entropy of the system will always increase.

Most of the systems we study in the laboratory are not *isolated* systems. In fact, a strictly isolated system does not exist. However, such a system is a convenient starting point for the development of a thermodynamic theory.

In a typical experiment carried out in a laboratory we usually keep the temperature and pressure constant, rather than the energy and the volume.

There is a well-established mathematical procedure to change variables, say from E, V, N to T, V, N, using the Legendre transformation (this is the same transformation used in classical mechanics to move from the Lagrange function to the more useful Hamilton function). We shall not discuss this mathematical procedure here. An excellent discussion of the Legendre transformation in thermodynamics is available in Callen (1985), and in classical mechanics in Goldstein, Poole, and Safko (2002).

We shall discuss here only the result of this transformation for two cases:

(a) Changing from the variables (E, V, N) to (T, V, N). One introduces a new thermodynamic function, called the Helmholtz energy, which is a function of the variables T, V, N and is defined by $A = E - TS$.

Where E is the internal energy, T the absolute temperature, and S the entropy of the system. In terms of the function $A(T, V, N)$, the Second Law states that removing a constraint in a constrained equilibrium state of a system characterized by the variables T, V, N will result in the Helmholtz energy always *decreasing*.

(b) Changing from (T, V, N) to (T, P, N). A further application of the Legendre transformation involves the replacement of the volume (V) by the pressure (P). In this case, one introduces a new function of the independent variables (T, P, N) called Gibbs energy, which is defined by $G = A + PV$.

In terms of Gibbs energy, the Second Law states that removing a constraint in a constrained equilibrium state of a system characterized by the variables T, P, N will result in the Gibbs energy always *decreasing*.

We shall further discuss the transformation from one set of variables to another in Section 1.6.

1.4 The Definition of Entropy Based on the Shannon's Measure of Information

1.4.1 *Introduction*

This is a relatively recent definition of entropy. It is, however, superior to both the Clausius and the Boltzmann definitions. Unlike the Clausius definition, which provides only a definition of changes in entropy, the present one provides the *entropy function* itself. Unlike the Boltzmann's definition which is strictly valid for isolated systems, and does not provide a simple intuitive interpretation, the present one is more general and provides a clear, simple, and intuitive interpretation of entropy. It is more general in the sense that it relates the entropy to probability distributions, rather than to the number of microstates. One final "bonus" afforded by this definition of entropy is that it not only removes any traces of mystery associated with entropy, but it also expunges the so-called irreversibility paradox.

This section will be longer than the previous sections. The reason is that most scientists are unfamiliar with the Shannon's measure of information (SMI). It is my conviction that any educated person should be exposed to the SMI, not because it is related to entropy, but because of its extreme generality, usefulness, and beauty.

In this section, we outline the procedure for building up the entropy function from the SMI. The SMI is a very useful concept. It was found useful in many branches of science. Here, we present only the qualitative aspects of the concept of the SMI, present a few simple examples, then describe the four steps leading from the SMI to the thermodynamic entropy.

Having done with the interpretation of entropy, i.e. answering the question: "What is entropy?" we turn to the meaning of the Second Law, which essentially answers the question: Why does entropy increase in a spontaneous process occurring in an isolated system? We also show the intimate relationship between the "what" and the "why" questions. Finally, we discuss the extension of these relationships to systems other than isolated ones.

A caveat

I said in the beginning that the present definition is recent. The unwary reader might be puzzled by this statement. Shannon himself referred to the SMI as entropy in 1948, so what makes this definition "recent"?

In this section, we shall make a clear-cut distinction between the SMI (sometimes referred to as Shannon entropy, or informational entropy), and the thermodynamic entropy. The fact that Shannon chooses to call his measure "entropy" was a serious mistake, which has caused great confusion in both information theory and in thermodynamics. This is explained in the following sections.

Furthermore, the application of information theory to statistical mechanics was pioneered by Jaynes (1957a and 1957b, 1965) and Katz (1967). Both authors used the so-called principle of *maximum entropy* to obtain the equilibrium distributions of a thermodynamic system. They then used these distributions in the Boltzmann's and Gibbs' definitions of entropy. In the present section, we use *both* the SMI as well as the principle of *maximum SMI* to obtain the entropy of an ideal gas, without recourse to either Boltzmann's or Gibbs' definitions. The first derivation of the *entropy function* for an ideal gas was published in Ben-Naim (2006b). It was later extended to include also the system of interacting particles in Ben-Naim (2008).

1.4.2 *What is the SMI?*

To understand entropy, one must understand the SMI, and to understand what the SMI is, one needs to know what a probability distribution is.

Consider any experiment, a game, or any set of events, such that each outcome is assigned a probability of occurrence, and the sum of all these probabilities is one (which means that one of these outcomes has occurred, or will occur).

A simple example is throwing a die. There are six outcomes: 1, 2, 3, 4, 5, and 6. Suppose we are given the probabilities of all these outcomes: p_1, p_2, p_3, p_4, p_5 and p_6. This set of probabilities is referred to as the probability distribution.[13]

Once we have the number of outcomes (or events, or possibilities, see examples below), and the corresponding probability distribution, we can define the SMI. We shall use the notation of Shannon for the SMI and write

$$H = -\sum_{i=1}^{n} p_i \log p_i$$

The sum (Σ) is over all n possibilities, and the symbol log, stands for the logarithm to the base 2. Here, we do not add a constant k as in the original definition of Shannon.

Examples

Example 1: Tossing a coin

Here, we have two possible outcomes (we always assume that one of these outcomes occurs, and we neglect the possibility that the coin will fall on its edge, or completely disappear). Let p be the probability of the outcome Head (H), and $(1-p)$ the probability of the outcome Tail (T). Then, it is easy to see that the SMI has the form shown in Figure 1.5.[14]

Note that the range of value of p is from zero to one, i.e. p must be a real number between zero and one ($0 \leq p \leq 1$). The range of values of the SMI, for this case, is between zero and one. Note also that the SMI is always positive and it has a *single* maximum value at $p = 1/2$, which is SMI $= 1$.

A quick qualitative meaning of the curve of SMI is shown in Figure 1.5. Suppose I toss a coin and you have to *guess* the outcome. I tell you that there are only two possibilities: H and T. Also, I tell you that the distribution is

case (a): $p_H = 1, p_T = 0$
case (b): $p_H = 0.9, p_T = 0.1$
case (c): $p_H = 0.5, p_T = 0.5$

In which case will it be more difficult to guess the outcome which has occurred?

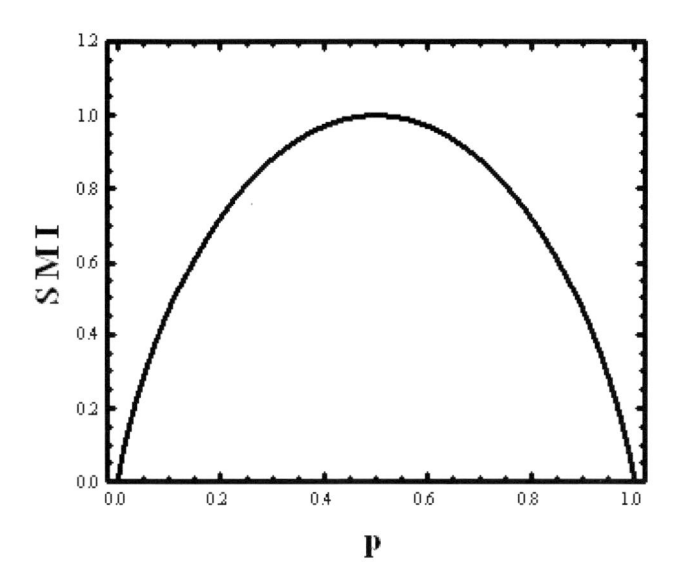

[Figure: SMI vs p, inverted-U curve peaking at p=0.5]

Fig. 1.5. The Shannon measure of information for the two-outcome experiment.

Suppose you are not told anything about the distribution, could you make any intelligent guess about the outcome?

You should realize that by being given the distribution, you were given some *information* about the outcomes. Can you order the three cases above, (a), (b), and (c), according to the extent of information you were given?

You should think about the answer before looking at Note 15.

I hope that the reader will realize that the amount of information provided by the distribution $P_H = P_T = 0.5$, is minimal. Equivalently, finding out which outcome occurred provides us with maximal information. This amount of information is called a *bit*. Although this book is not about information theory, it should be noted that there are two meanings to the *bit*. The first is for a binary digit, which is either zero or one in a binary number such as 11010011. The second is a *unit of information*. The *bit* in this case is the maximum amount of information one gets for a binary question, when the two events have equal probability.

Example 2: Throwing a die

In this case, we have six outcomes. Suppose I give you the probabilities of the outcomes:

case (a): $p_1 = 1, p_2 = 0, p_3 = 0, p_4 = 0, p_5 = 0, p_6 = 0$
case (b): $p_1 = 0.8, p_2 = 0.1, p_3 = 0.1, p_4 = 0, p_5 = 0, p_6 = 0$
case (c): $p_1 = \frac{1}{6}, p_2 = \frac{1}{6}, p_3 = \frac{1}{6}, p_4 = \frac{1}{6}, p_5 = \frac{1}{6}, p_6 = \frac{1}{6}$

I throw the die in each case. You are informed about the number of outcomes and their probabilities. You do not know which outcome has occurred.

You are offered to ask binary questions, i.e. questions which can be answered by Yes, or No. You have to pay one dollar for each answer you get, and receive ten dollars once you find out which outcome occurred.

Which game will you choose to play: (a), (b), or (c)? Can you estimate the minimum number of questions you need to ask in order to get the information about which outcome occurred? I urge you to think about the answers to these two questions before looking at Note 16.

1.4.3 Uniform and Non-Uniform Distribution

Here we present the general definition of the SMI which is the cornerstone of information theory. We shall present the SMI in a qualitative manner just enough to understand how entropy may be interpreted in terms of the SMI. We shall proceed to develop this concept in the following three steps.

First, we start by playing a simple and familiar 20-question (20Q) game. I choose an object out of n possible objects, and you have to find out which object I have chosen by asking binary questions, i.e. questions which are answerable by either Yes or No. We assume in this game that the probability of choosing the object out of n objects is the same for each object and is equal to $1/n$. We will refer to this game as the *uniform probability game*, or simply the uniform game. To

The Various Definitions of Entropy | 33

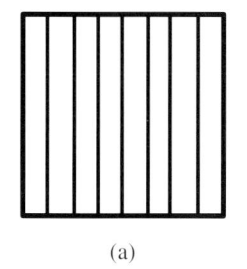

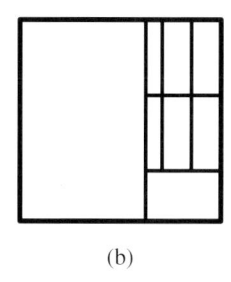

Fig. 1.6. (a) A uniform, and (b) a non-uniform game having the same number of outcomes.

make the game more precise and also to ease the generalization to the non-uniform game, think of the game shown in Figure 1.6a. You are shown a board which is divided into n regions of equal areas. You are told that a dart was thrown by someone in a blindfold, and hit a point on the board. You have to find out in which region the dart hit the board by asking binary questions. I trust that you can easily play this game.

Second, we generalize the simple 20Q game described above to a more complicated game. The objects are chosen with different probabilities. To understand this more general game, consider a board on which a dart was thrown. The board is divided into n regions of *unequal* areas, and the dart was thrown by someone in a blindfold. Now you are told that the dart did hit a point on the board. You are also informed about the relative areas of all the regions in Figure 1.6b. Your task now is to find out in which area of the board the dart hit by asking binary questions. You should realize that this game is different from the previous game because the "events" here are not equally probable. We refer to this game as the *non-uniform game*.

A quick question before we proceed. Suppose you are offered to play the 20Q game on either the board of Figure 1.6a or 1.6b. You have to pay one dollar for each question you ask. Once you find out where the dart hit, you get a prize of 20 dollars. Which game would you choose to play?

Now, we turn to the third, and the last, step. If you understand the 20Q game, and if you can answer my question above (about the preferable game out of the two in Figures 1.6a and 1.6b), you should realize that in playing the 20Q game, you *lack* information on the location of the dart. By asking binary questions you *gain* information from each answer you get (and you have to pay to get that information). Eventually, you will obtain all the information you need.

The last step does not introduce any new *generalization* to the 20Q game. It only makes the same game very large — very much larger than the games you are used to playing, or games shown in Figures 1.6a and 1.6b. The game is large indeed, but there is nothing new in principle. You cannot possibly play this game because you do not have enough time in your life to ask so many questions — but you can at least *imagine* playing such a game. It will perhaps be called the 10^{23} Q game, rather than the 20Q game, but you can be assured that if you only imagine playing this game, you will understand the concept which is still considered by many as the most mysterious concept in physics — the entropy. We shall discuss this in Section 1.4.5.

We now elaborate on each of these three steps before we define entropy.

First step: Playing the uniform 20Q game

This is the relatively easy game. A dart hit a board which is divided into n equal-area regions. You are told that the dart was thrown by someone blindfolded, and it hit one region of the board. You also know that the probability of hitting any one of the regions is $1/n$. This is why we refer to this game as the *uniform* game.

For $n = 8$, how many questions do you need in order to find out where the dart is? If you are smart enough, you can obtain the information on where the dart is in three questions. How do you do this?

You simply divide the eight regions into two groups, each containing four regions, and ask, "Is the dart in the right group?" If the answer is "Yes," you divide the remaining four regions on the right into two groups, and ask, "Is it in the right group? and so on." In this method you will find out where the dart is in exactly three questions. This way of asking questions is referred to as the *smartest* way, or the smartest strategy. You can check and convince yourself that if you use any other strategy of asking questions, you will need on average, to ask many more questions. One can prove mathematically that by asking questions of this kind (i.e. dividing each time into two equally probable regions) you get *maximum information* for each answer. Therefore, you will need to ask a minimum number of questions to get the information you need. Figure 1.7 shows the two strategies of asking questions for the case $n = 8$. Note that the amount of information (measured in bits) is independent of the strategy one chooses in asking questions. If you ask smart questions, you get the same amount of information in the smallest number of questions. For more details, see Ben-Naim (2012, 2015a).

Take note that the number of questions (here, 3) is related to the number of regions (8) by the logarithm to the base 2. In this case: $3 = \log_2 8$ (here, the base of the logarithm is 2). You can check for yourself

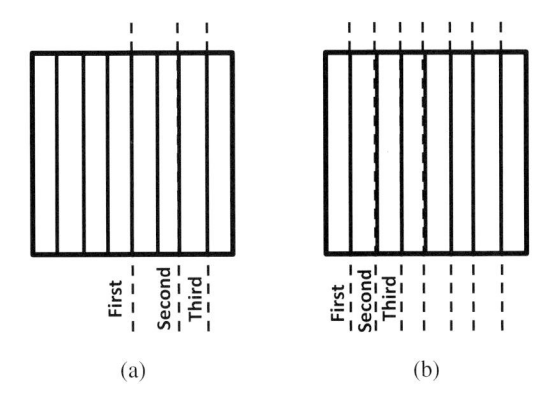

Fig. 1.7. Two strategies for asking questions. The dashed lines show the division of the total number of possibilities into two parts: (a) the smartest, and (b) the dumbest strategy.

that for $n = 16$ regions, you will need four questions. For $n = 32$, you will need five questions, and for $n = 2^k$ with an integer k, you will need k questions to ask. Take note that when we *double* the number of regions, the number of questions you need to ask (smart questions) increases by *one*. One can prove in general, that for any n, the average number of questions you need to ask is approximately $\log_2 n$. We say "average," because for any n, say $n = 7$, you cannot divide at each step the total number of regions into two equally probable groups. But you can try to do it as close as possible, for instance divide the 7 regions into 4 and 3 regions. The general result for any number n of equally probable regions is that you need to ask, on the average, about $\log_2 n$ questions in order to obtain the required information. All this is valid for a system of equally probable regions. It is very easy to prove that the relationship between the number of regions and the number of questions you need to ask is given by $H \approx \log_2 n$. Figure 1.8 shows the average number of questions one needs to ask as a function of n, for the *smartest* and for the *dumbest* strategies. Details of how these curves were calculated are given in Ben-Naim (2008, 2016a).

Second step: Playing the non-uniform 20Q game

By "non-uniform game" we mean unequal probabilities to the different events. To be specific, consider the modified game shown in Figure 1.6b. We have again a board divided into eight regions. But unlike in the game in Figure 1.6a, the areas (hence the probabilities) of the eight regions are different.

Usually, when we play the parlor game of 20 questions we assume implicitly that the probability of choosing a specific object or a specific person is $1/n$, where n is the total number of objects from which one is selected.

In the example shown in Figure 1.6b, the areas of the regions are not equal. This makes the *calculation* of the number of questions more *difficult*, yet the actual *playing* of this game is *easier* than the one shown in Figure 1.6a.

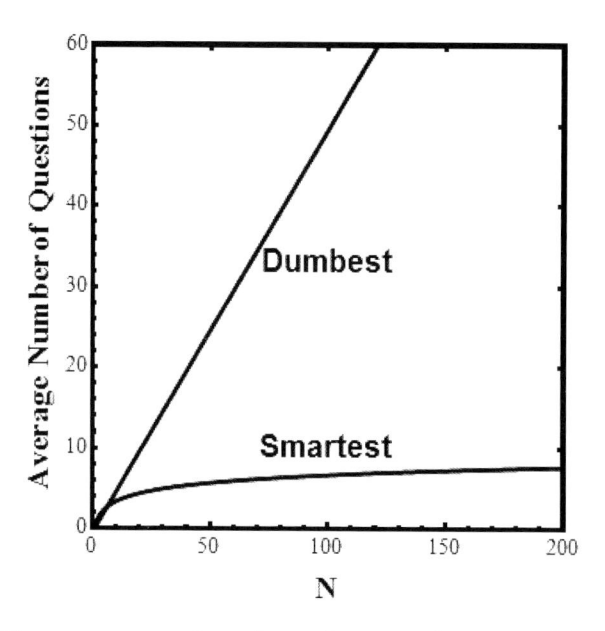

[Graph showing Average Number of Questions vs N, with two curves labeled "Dumbest" (linear increasing) and "Smartest" (logarithmic).]

Fig. 1.8. The average number of questions as a function of the number of objects in a 20Q game, or outcomes in an experiment.

One can prove mathematically that the average number of (smart) questions you need to ask in order to obtain the information (on where the dart is) is given by the Shannon formula $H = -\sum p_i \log_2 p_i$ where p_i is the probability of the event i, which in our case is simply the relative area of the region i. We shall refer to H as the Shannon measure on information (SMI).

One can also prove that the quantity H defined above for a non-uniform distribution is *always smaller* than the quantity H defined on a uniform distribution with the same total number n. In terms of the 20Q game we can say that playing the game of Figure 1.6b is always *easier* than playing the game in Figure 1.6a. Easier in the sense that you will need, on average, fewer questions to ask in order to obtain the information you need.

This mathematical result is intuitively quite obvious. If you are offered to play either one of the two games in Figures 1.6a and 1.6b, and you have to pay for each answer you get, it is always advantageous

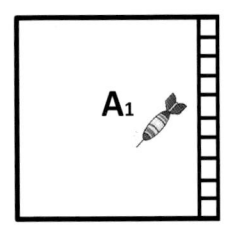

Fig. 1.9. An extreme non-uniform game.

to play the non-uniform game. You can try to play the two games in Figure 1.6 and convince yourself that the uniform game always requires more questions to ask. This answers the question we posed at the end of Section 1.4.2. If you are not convinced, try to play a more extreme game shown in Figure 1.9. Here, if you know the distribution, you should ask the first question: "Is the dart in area A_1?" The chances to get a "Yes" answer are 9/10, and to get a "No" are 1/10. This means that if you play the game many times, you will get the information (on where the dart is), in about one question. In fact, from Shannon's formula you will find that the average number of questions is less than one. This means that even without asking any questions, you will know with high probability where the dart is. For more details, see Ben-Naim (2008).

Third step: Generalization for a 20Q game to an over 20^{23}Q game

Now that you have an idea about the relationship between the number of objects, and the average number of binary questions you need to ask in order to find out which one of the objects was selected, let us play a very large game of the same kind.

Suppose you have a single atom in a cubic box of edge d. The *state* of the atom can be described by its location and its velocity at each instance of time. Let l be its location and v its velocity. Clearly, the pair (l, v) describes the state of this atom. It is also clear that there are infinite states of such kind. Therefore, if I know the *exact* state of the atom, and you have to find its state by asking binary questions, you will need on average, an *infinite* number of questions to ask.

The Various Definitions of Entropy | 39

Fortunately, there is the uncertainty principle in physics. This principle states that you cannot determine the exact location and velocity of the atom, but there is a limit to the "size" of the box ($\Delta l \Delta v$) in which you can determine the state of the atom. This passage from the infinite number of states in the *continuous* range of locations and velocities, to the finite number of possibilities is shown schematically in Figure 1.10. Here, we reduce the *infinite* number of points in the range (0, 1) to a *finite* number of small intervals.

Now, if I know the state of the atom and you have to ask binary questions, you will need only a finite number of questions to ask. This game is no different from playing the 20Q game that you are familiar with. Next, we move from one atom to a huge number of atoms, say 10^{23} atoms in the same box of edge d. The problem is now to find the "state" of this huge number of atoms — not the exact state, but an approximate state as is imposed by the uncertainty principle. Figure 1.11 shows a configuration of ten particles. Note that each

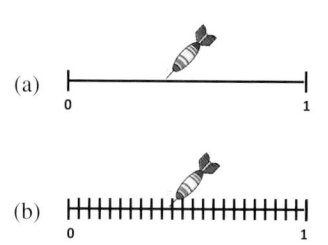

Fig. 1.10. Passage from the infinite (continuum) (a), to the discrete (b) description of the states in a one-dimensional system.

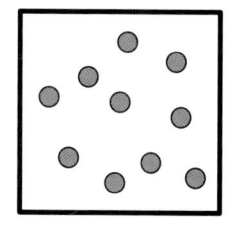

Fig. 1.11. A specific (locational) configuration of ten particles.

particle's state is characterized by its locational and velocity vector, as shown in Figure 1.4. Again, there is no difficulty in playing this game. You will need to ask many questions, far too many than you can achieve in your lifetime, or during the whole age of the universe. However, there is no difficulty *in principle* in imagining playing such a huge game. There will be a finite number of questions — finite, albeit a huge number.

Finally, we need to introduce one principle from physics to reach for the entropy. The particles are *indistinguishable*. This means that if you exchange the locations of two atoms, you get the same configuration. Figure 1.12 illustrates this reduction in the number of configurations for three particles. As can be seen on the left-hand side, there are six *different* configurations. The six configurations coalesce to one configuration when the particles are indistinguishable.

The conclusion is that whenever we have a finite number of outcomes (or events, states, configurations, etc.), and their probability distribution $p_1, \ldots, p_n$ is given, we can define the corresponding SMI by the general definition given in Box 1.1. For a given n, the maximal value of the SMI (H_{max}) will be obtained for the uniform distribution, i.e. when $p_i = 1/n$ for each i, in which case $H_{max} = \log n$.

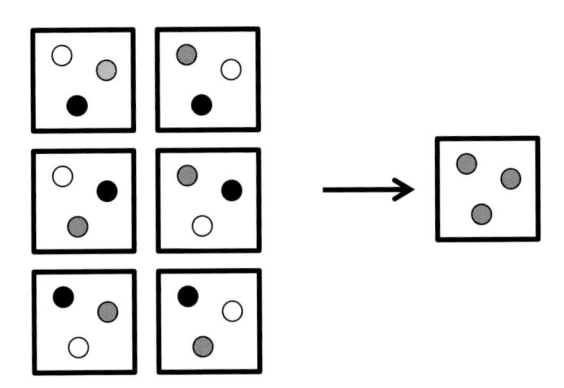

Fig. 1.12. The reduction of the number of configurations when the particles become indistinguishable.

The Various Definitions of Entropy | 41

$$H = -\sum_{i=1}^{n} p_i \log p_i$$

$$H_{max} = \log n$$

Box 1.1

This can be proven mathematically [see Ben-Naim (2008, 2012, 2015a)].

In terms of the 20Q game we can say that the game with uniform distribution is the most difficult game, given the total number of outcomes n. Again, here "difficult" means that we need to ask, on average, more questions to find out which outcome has occurred.

We are now ready to derive the entropy of an ideal gas from the SMI. Before doing this, let us pause to discuss the different interpretations of the SMI.

1.4.4 *Interpretation of the SMI*

Here we discuss three interpretations of the SMI: the first, as an average of the uncertainty about the outcomes of an experiment, the second, as a measure of the average unlikelihood, and the third as a measure of information. We shall use the letter H for the quantity defined above, and refer to it simply as the SMI. It should be noted again that the SMI is a measure of information only in a very restricted sense. The meaning, the value, the importance, or the content of the information is irrelevant to the SMI.

The uncertainty meaning of the SMI

The interpretation of H as an average *uncertainty* is very popular. This interpretation is derived directly from the meaning of the probability distribution.

Suppose we have an experiment having n possible outcomes with probability distribution $p_1, \ldots, p_n$. If we know that $p_1 \approx 1$, and all others are $p_i \approx 0$, we are *certain* that the outcome 1 occurred or will occur. For any other value of p_i we are *less certain* about the occurrence of the event i. *Less certainty* can be translated into *more uncertainty*. Therefore, the larger the value of $-\log p_i$, the larger the uncertainty about the occurrence of the event i. Multiplying $-\log p_i$ by p_i, and summing over all i, we get an *average* uncertainty about *all* the possible outcomes of the experiment. See Note 3 to the Preface.

The unlikelihood interpretation

A slightly different but still useful interpretation of H is in terms of *likelihood* or *expectedness*. These two interpretations are also derived from the meaning of probability. When p_i is small, the event i is unlikely to occur, or its occurrence is less expected. When p_i approaches one, the event i becomes more likely to occur, or is more expected. Since $\log p_i$ is a monotonic increasing function of p_i, we can say that the larger the $\log p_i$, the larger the likelihood or the larger the expectedness of the event i. Since $0 \leq p_i \leq 1$, we have $-\infty \leq \log p_i \leq 0$. The quantity $-\log p_i$ is thus a measure of the *unlikelihood* or the *unexpectedness* of the event i. Therefore, the quantity $H = -\sum p_i \log p_i$ is a measure of the *average unlikelihood*, or *unexpectedness*, of the entire set of the events.

The meaning of the SMI as a measure of information

As we have seen, both the uncertainty and the unlikelihood interpretation of H are derived from the meaning of the probabilities p_i. The interpretation of H as a measure of information is a little trickier and less straightforward. It is also more interesting since it conveys a different perspective on the Shannon measure of information. It should be emphasized from the outset, that the SMI is not information. It is also not a measure of any type of information, but of a very particular kind of information. The confusion of the SMI with information is almost the rule, not the exception, by both scientists as well as non-scientists. See also Ben-Naim (2015a).

Some authors assign to the quantity $-\log p_i$ the meaning of information (or self-information) associated with the event i, then interpret

H as the *average* information about the outcomes. This is not a correct interpretation of the SMI, for the following reasons.

The idea that one gains more information from the occurrence of a rare event is incorrect. This concept emerges erroneously from rationalizing that because the $-\log p_i$ becomes large for events with small p_i, somehow more information has been gained. When I know that an event i has occurred, I got the information about the occurrence of the event i. I might be surprised to learn that a rare event has occurred, but the size of the information one gets when the event i occurs is not dependent on the probability of that event.

When an experiment is performed and I am informed that the outcome, say "k," occurred, the information I have, is

$$\text{Outcome "}k\text{" occurred}$$

The Shannon measure of this information depends on the number of letters and the probability distribution (or frequencies) of the letters of the alphabet in the particular language this information is conveyed. Shannon was interested in such as a measure of information. This measure can be written as $-M \sum_{i=1}^{n} q_i \log q_i$, where M is the number of letters in the message, and n is the number of letters in the alphabet of the language, and q_i is the frequency of the letter i in that particular language.

Pause and ponder

You are given a coin with a probability distribution

$$\Pr(\text{Head}) = 1$$
$$\Pr(\text{Tail}) = 0$$

Consider the following two cases:

(a) I tell you that I tossed the coin and the outcome "Head" occurred. Answer the following questions:

  1. What was the information that I conveyed to you?
  2. How much information did you get from me?

(b) I tell you that I tossed the coin and the outcome "Tail" occurred.

Answer the same questions above (1 and 2), for this case. My answers are given at the end of Section 1.4.4.

Both p_i and $\log p_i$ are measures of the uncertainty about the occurrence of an event. They do not measure *information* about the events. Therefore, $-\log p_i$ is not the information (or self-information) associated with the event i. Hence H cannot be interpreted as an average information associated with the entire experiment. Instead, we assign "informational" meaning directly to the quantity H, rather than to the individual events.

As we have seen in the examples of the 20Q games, the value of H measures the *size* of the game in the following sense. An experiment is performed and we have to find out which event has occurred by asking binary questions. A "bigger" game means that we need to ask more questions in order to get the information on which event has occurred.

It should be stressed that the SMI provides the average number of (smart) questions one needs to ask, in order to obtain the information. The *value* of the SMI is an objective quantity which "belongs" to the *entire* game, not to individual outcomes. Knowing n and the probability distribution of outcomes determines the value of the SMI. It does not matter who chose the object in the game, or who has to find out which object was chosen, or which outcome has occurred. The value of the SMI is the same for anyone who is given the probability distribution. This comment will be important when we come to interpret the meaning of entropy.

A caveat

A statement like "SMI is uncertainty" is true, but not the whole truth. *One must specify uncertainty with respect to what, and one must also show how this uncertainty follows from the definition.*

Here is an example. We have a die; we might have uncertainties about its color, its weight, the material it is made of, the outcome when we throw it, and many more. All these uncertainties have nothing to do with either the SMI or entropy. There is one uncertainty which is associated with the *distribution* of the results, and which is related to the SMI of the outcomes. Here is the tricky point. The SMI is not the uncertainty about the outcome of the result, i.e. if I throw a die and ask you what the result is, you can say I am not certain about the result, or I have great uncertainty about the result. This is a valid answer, but it is not the SMI. The SMI is the average uncertainty about *all the results, when you know the probability of the outcomes.*

Once you know the distribution, and I throw the die, the average uncertainty regarding all the outcomes is the SMI. It is defined for the *whole probability distribution*, and it does not depend on whether one knows or does not know the result. If you do not know the distribution, then, of course, you are uncertain about the outcome, as well as about the distribution. If you know the distribution, the uncertainty, as measured by the SMI, depends on the distribution; it could be zero, when one event is certain to occur, and maximal uncertainty when the distribution is uniform. Here we discussed the uncertainty meaning of the SMI, not of entropy. We shall discuss this interpretation of entropy in Section 2.12.

Answers to the questions

Case (a): 1. The information is: "Head has occurred."
2. The size of this information depends on which measure of information you are using.
Case (b): 1. The information is: "Tail has occurred."
2. The same answer as in case (a).

1.4.5 Applying the SMI to Obtain the Entropy of an Ideal Gas

Now that we know what the SMI is, and how to calculate the SMI for any given distribution (of n events), we turn to the application of the SMI to a simple thermodynamic system.

Consider N simple particles, say argon atoms, in a box of volume V, and having some fixed energy E. This triplet of variables (E, V, N) defines the *macroscopic state* of the system. We also assume that the particles have no internal structure (say, hard spheres), and the system is very dilute so that the interaction energy between the particles is negligible. In this case, all the energy of the gas is due to the kinetic energy of the particles.

In such a system, the temperature is related to the average kinetic energy of the particles. The higher the temperature, the more "energetic" the particles are.[17]

Clearly, for any given macroscopic state (E, V, N), there are many *microscopic* states. Figure 1.13 shows a system of ten particles at three different *configurations*. By configuration we mean here a specification of the location and velocity of each atom in the system, as depicted in Figure 1.4.

When we say that the thermodynamic system is well defined by the variables E, V, N, we mean that the total number of particles N is fixed. Each particle can access the entire volume V, and the average kinetic energy of all the particles is E. Such a system is referred to as isolated system.

Clearly, there are many microscopic configurations which are consistent with the requirement that E, V, N are constants. In fact, there is an infinite number of such microscopic configurations. However, for all practical purposes we can assume that both the location and the velocity of each particle are specified to within a small "box," say, the location of the particles is within a box of volume $dV = dxdydz$,

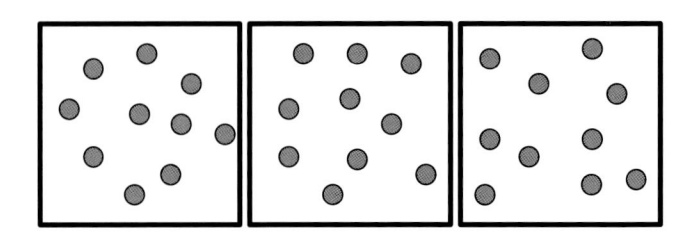

Fig. 1.13. Three configurations (microstates) of the same system.

The Various Definitions of Entropy | 47

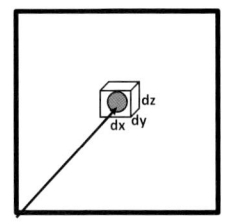

Fig. 1.14. The location of each particle is characterized within a small box of volume $dV = dxdydz$ and a small box of "volume" $dv_x dv_y dv_z$.

where dx, dy, and dz are the edges of an infinitesimal box of size dV (see Figure 1.14). Similarly, we can specify the velocity of the particles within a small "box" of velocities of size $dv_x dv_y dv_z$.[18]

The road leading from the SMI to entropy is highly mathematical. We shall outline here only the general steps, in a qualitative way. More details on the procedure is available in Ben-Naim (2008, 2012, 2015a).

The overall plan of obtaining the entropy of an ideal gas from the SMI consists of four steps. First, we calculate the locational SMI associated with the equilibrium distribution of locations of all the particles in the system. Second, we calculate the velocity SMI associated with the equilibrium distribution of velocities (or momenta) of all the particles. Third, we add a correction term due to the quantum mechanical uncertainty principle. Fourth, we add a correction term due to the fact that the particles are indistinguishable.

Note that in the first two steps, we use the maximum SMI method to find out the equilibrium distribution. Then we use this equilibrium distribution to evaluate the corresponding SMI. In the last two steps we introduce two corrections to the SMI. Once we combine the results of the four steps, we get, up to a multiplicative constant, the entropy of an ideal gas.

(i) The locational SMI of an ideal gas

Consider first the case of one particle which freely moves in a one-dimensional (1-D) "box" of length L. Clearly, there are infinite numbers of points in which the center of the particle may be found. However, we are never interested in the *exact* point in which the particle is, but rather in which small interval dx the particle is. The formal problem we pose is to find the maximum value of the SMI, subject only to the normalization condition. This means that the probability of finding the particle anywhere within the box of length L is one. The result is given in Box 1.2. Note that $f(x)dx$ is the probability of finding the center of a particle between x and $x + dx$. $f^*(x)dx$ has the same meaning but at equilibrium.

In this case the probability density $f^*(x)$ is constant, independent of x. Also note that the distribution $f^*(x)$ which maximizes the SMI is also the actual experimental distribution at equilibrium. The corresponding locational SMI (at equilibrium) is given in Box 1.2.

We shall refer to the density function $f^*(x)$ as either the function that maximizes H, or as the equilibrium density function. The latter interpretation will be clear when we interpret the final result of H as the entropy of an ideal gas at equilibrium.

We note here that $\log L$ is to be understood as the SMI associated with the location of a particle in the 1-D system of length L. The larger the L, the larger is the SMI. One should keep in mind that the

---

The equilibrium distribution of locations of one particle in a one dimensional system: $f^*(x) = 1/L$

$H_{max}(location\ of\ one\ particle\ along\ the\ x-axis) = \log L$

The equilibrium distribution of locations of one particle in a 3D system of volume V: $f^*(x, y, z) = 1/V$

The corresponding SMI is: $\log V$

The SMI for N independent and distinguishable particles is

$H_{max}^D(R^N) = NH_{max}(x, y, z) = N \log V$

Box 1.2

SMI for the continuous case is actually a divergent quantity. In all the application of this equation, we either make a *discretization* of the length L, or take the *difference* of SMI for the two values of L. In both cases we remove any problem arising from the divergence of the SMI. [For details, see Ben-Naim (2008)].

Note that log stands for the logarithm to the base 2. We shall treat L as a pure number (i.e. dimensionless). In fact, this number is L/dx, the number of intervals of length dx. We shall ignore the units of length, as well as the units of any other quantity under the logarithm sign. In the final expression, we must have a pure number under the logarithm.

The generalization of this result to the three-dimensional (3-D) case is straightforward. Suppose the particle is confined to a cubic box of edge L and volume $V = L^3$. Clearly, the SMI associated with the y-axis and the z-axis will be the same as the SMI associated with the x-axis. Furthermore, we assume that the events "being at a location x," "being at a location y," and "being at a location z" are independent events. Therefore, the SMI associated with the location x, y, z within the cube of volume V is the sum of the SMI associated with the three axes. Thus, if we use the shorthand notation $H_{max}(x)$ for the SMI in a 1-D system, we can write the locational SMI of a particle in a 3-D system as the sum of the SMI associated with each axis (see Box 1.2).

We next extend this result to the case of N-*independent* and *distinguishable* (D) particles. We also use the short hand notation: $R_i = (x_i, y_i, z_i)$ for the locational vector of particle i, and $R^N = R_1, \ldots, R_N$ for the locational vector of all the N particles.

Since the particles are assumed to be independent, the SMI of the N particles is simply the sum of the SMI of all the single particles. The SMI for N, independent, and distinguishable particles is given in Box 1.2.

Note that we added the superscript D for distinguishable particles. We shall soon see that the fact that the particles are indistinguishable (ID) introduces a correlation between the particles which causes a reduction in the SMI of N particles. Note also that we still retain the subscript "max." This will be important when we identify the maximal value of H with the entropy of the system at equilibrium.

(ii) The momentum SMI of an ideal gas

Again, we start with particles moving along the 1-D system of length L. We are interested in the probability density of finding a specific particle with velocity between v_x and $v_x = v_x + dv_x$. We assume that the particles can have any value of v_x from $-\infty$ to $+\infty$, but we require that the average kinetic energy of the particles is constant. We shall skip the details of the derivation. The solution to this problem is given in Box 1.3. We shall refer to $f^*(v_x)$ as the equilibrium density distribution of the velocities in one dimension.

Figure 1.15 shows the distribution $f^*(v_x)$ for various values of T (for this illustration we take $m = 1, k_B = 1$). We see that the larger the temperature, the larger the spread of the distribution of the velocities. Thus, the average "width" of the distribution may be described either by the variance σ^2 or by the temperature T.[19]

In terms of the temperature, we rewrite the SMI associated with the equilibrium distribution $f^*(v_x)$ is given in Box 1.3. The meaning of this quantity is derived from the meaning of the SMI. Note again that

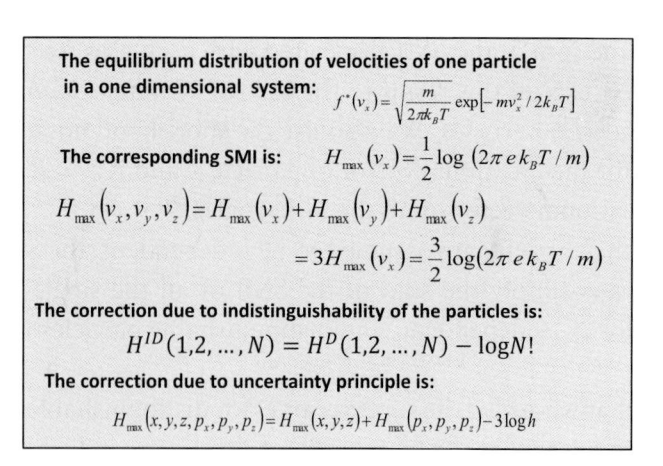

The equilibrium distribution of velocities of one particle in a one dimensional system:
$$f^*(v_x) = \sqrt{\frac{m}{2\pi k_B T}} \exp\left[-mv_x^2 / 2k_B T\right]$$

The corresponding SMI is: $H_{\max}(v_x) = \frac{1}{2} \log(2\pi e k_B T / m)$

$$H_{\max}(v_x, v_y, v_z) = H_{\max}(v_x) + H_{\max}(v_y) + H_{\max}(v_z)$$
$$= 3 H_{\max}(v_x) = \frac{3}{2} \log(2\pi e k_B T / m)$$

The correction due to indistinguishability of the particles is:
$$H^{ID}(1,2,\ldots,N) = H^D(1,2,\ldots,N) - \log N!$$

The correction due to uncertainty principle is:
$$H_{\max}(x, y, z, p_x, p_y, p_z) = H_{\max}(x, y, z) + H_{\max}(p_x, p_y, p_z) - 3 \log h$$

Box 1.3

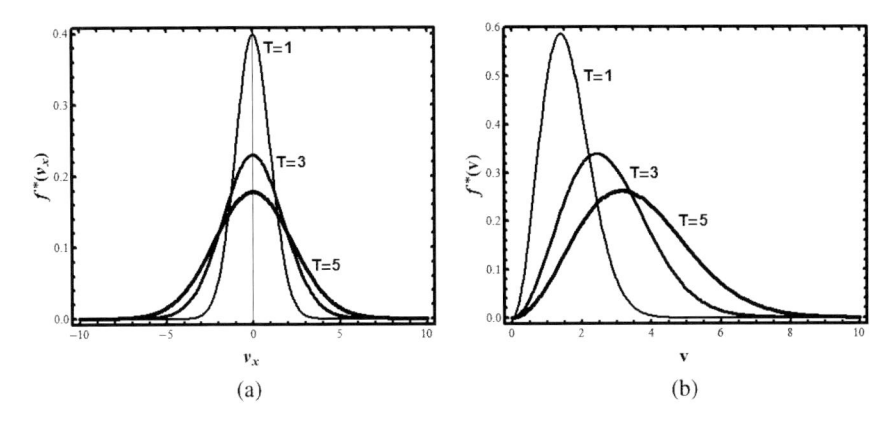

Fig. 1.15. (a) The velocity distribution of particles in one dimension at different temperatures, and (b) the speed (or the absolute velocity) distribution of particles in three dimensions at different temperatures.

the SMI of any continuous variable has a divergent part. However, in actual application of H_{max} we either make a discretization of the infinite range $(-\infty, \infty)$ into a finite number of small intervals, or we take differences in the values of H between two states; here, two temperatures. Doing this removes the divergent part of the SMI. The important result we have obtained is that the larger the temperature (or equivalently the average kinetic energy of the particles), the larger is the SMI or the uncertainty associated with the distribution of the velocities. We note again that at this stage, we ignore the units in the expression under the logarithm.

We next assume that the velocities along the three axes v_x, v_y, v_z are independent. Therefore, the SMI for a single particle moving with velocities v_x, v_y, v_z is given by the sum of the SMI for each axis (see Box 1.3).

For the purpose of constructing the entropy of ideal gas, we shall need the distribution of the momenta. This is simply obtained by the transformation $p_x = mv_x, p_y = mv_y, p_z = mv_z$. We shall skip the details here. The interested reader can find the mathematical details in Ben-Naim (2008, 2012).

(iii) A correction due to the indistinguishability of the particles

We have already seen in Figure 1.12 that indistinguishability of the particles reduces the total number of configurations of the system. This can be cast in the form of mutual information.

This term is a measure of the correlation between different information. In Box 1.2 we wrote the SMI of N particles as a sum of the SMI of all the particles presuming that they are independent. For the case of N indistinguishable particles, we have to add a correction term which is given in Box 1.3.

We conclude that the indistinguishability of the particles introduces a correlation between the particles, which causes a reduction of the SMI. We have calculated the mutual information of indistinguishable particles by calculating the change in the number of configurations of the system caused by "erasing the labels" on the particles. The reader should be convinced by checking a few examples that whenever we "un-label" the particles, the number of configurations or arrangements is always reduced, and as a result of which the value of the SMI of the system is reduced too.

(iv) A correction due to the uncertainty principle

In step (*i*), we calculated the locational SMI for a single particle. In step (*ii*) we calculated the velocity or the momentum SMI of a single particle. These are given in Boxes 1.2 and 1.3. We now wish to find out the SMI associated with *both* the location and the momentum.

Classical thinking would have led us to conclude that the SMI associated with both the location *and* momentum of a particle should be the sum of the two SMIs. However, quantum mechanics tells us that the accuracies in determining the location and the momentum of a particle are *not* independent. This is the well-known Heisenberg uncertainty principle. For our case the uncertainty principle states that we cannot determine *both* the location and the momentum within an accuracy of the order of h; here h is the Planck constant, $h = 6.626 \times 10^{-34}$ Js. Therefore, we must add a correction to account for the uncertainty principle.

Once we take into account the uncertainty principle, we get the SMI for one particle in one dimension. For the three-dimensional case we have to add the correction term $-3\log h$, i.e. we subtract $\log h$ for each degree of freedom.

Finally, for N indistinguishable and non-interacting particles, we have

$$H^{ID}(1, 2, \ldots, N) = H_{max}^{D}(\mathbf{R}^N) + H_{max}^{D}(\mathbf{p}^N) - \log N! - 3N \log h$$

This is an important result. To obtain the SMI of N particles, described by their locations and momenta, we first treat the particles as being distinguishable and classical. In this case, we can sum the SMI associated with all the locations ($\mathbf{R}^N$), and all the momenta ($\mathbf{p}^N$) of the particles. Then, we add two corrections, one due to the indistinguishability of the particles, and the other due to the Heisenberg uncertainty principle. These two corrections change the total SMI by the amount $\log N! + 3N \log h$. Note again that at this stage we consider h to be a pure number. We shall fix its units only in the final expression for the entropy.

(v) The entropy of a classical ideal gas

In Section 1.4.4 we calculated the maximal value of the SMI of a system of N simple, non-interacting and indistinguishable particles.

Recall that the SMI may be defined for *any* distribution. It can be defined for any distribution of locations and any distribution of momenta, not necessarily at equilibrium. It can be defined for any number of particles and can be defined for distinguishable or indistinguishable particles. All these have nothing to do with the entropy. Up to this point you can rightfully regard the SMI as a quantity that measures the size of a huge 20Q game about all the locations and the momenta of all the particles in the system.

Here we are interested in *very special distributions*. These are the distributions of locations and momenta that maximize the corresponding SMI. We denoted these special distributions with asterisks, and the corresponding SMI by H_{max}. However, we also know that starting with any arbitrary distribution of locations and momenta, the system will tend to a limiting equilibrium distribution, the uniform distribution for locations, and the normal distribution for the momenta. Therefore, we shall refer to the distribution that maximizes the SMI as the *equilibrium distribution*. We shall soon see that the distribution that maximizes the SMI is also the distribution which maximizes the probability.[20]

Here we make a huge conceptual leap, from SMI of 20Q games to a fundamental concept of thermodynamics. As we shall see soon, this leap is rendered possible by recognizing that the SMI associated with the equilibrium distribution of locations and momenta of a large number of indistinguishable particles is *identical* (up to a multiplicative constant) with the statistical mechanical entropy of an ideal gas. Since the statistical mechanical entropy of an ideal gas has the same properties as the thermodynamic entropy as defined by Clausius, we can declare that this special SMI is identical to the entropy of an ideal gas. This is a very remarkable achievement. Recall that von Neumann suggested naming the SMI, entropy. This was a mistake. In general, SMI is not entropy. Only when you apply the SMI to a *special* distribution does it become identical with entropy.

We recall that the SMI of a system of N particles at equilibrium has two contributions due to location and momentum, and two corrections due to the indistinguishability of the particles and the uncertainty principle. Thus, we have the following expression for the SMI of N non-interacting particles at equilibrium (see Box 1.4).

In order to obtain the expression for the entropy of an ideal gas, all we have to do is to use the natural logarithm and multiply H^{ID} by the Boltzmann constant k_B, i.e. $S = (k_B \ln 2)H$. Thus, we define the entropy of an ideal gas of simple particles as the maximal value of the SMI associated with the distribution of locations and momenta at equilibrium.

The Various Definitions of Entropy | 55

$$H^{ID}(1, 2, \cdots, N)$$
$$= H_{max}(locations) + H_{max}(momenta)$$
$$- I(uncertainty\ principle) - I(ind)$$
$$= N\log\left[\frac{V}{N}\left(\frac{2\pi m k_B T}{h^2}\right)^{3/2}\right] + \frac{5N}{2}$$

$$S(ideal\ gas) = Nk_B \ln\left[\frac{V}{N}\left(\frac{2\pi m k_B T}{h^2}\right)^{3/2}\right] + \frac{5Nk_B}{2}$$

Box 1.4

The multiplication by a constant k_B, as well as changing to the natural logarithm, determines the units in which we measure entropy. It does *not* affect the meaning of entropy, as the SMI associated with the location and the momenta of a system of N non-interacting particles at equilibrium. We normally apply this identity between entropy and SMI for a thermodynamic system, i.e., when N is very large, V is very large, but the density N/V is constant.

The expression for the entropy shown in Box 1.4 is identical with the equation obtained by Sackur and Tetrode in 1912, based on Boltzmann's definition of entropy. In Section 3.2, we shall compare the values of the entropies of some gases calculated using experimental data on one hand, and the theoretical values calculated from the various degrees of freedom of the molecules, on the other hand. In the case of argon the main contribution to the entropy is the translational entropy, which is the entropy that can be calculated from the Boltzmann equation, the Sackur-Tetrode equation, or from the SMI. In many cases there is an excellent agreement between the theoretical values and the experimental values. It should be noted that in many cases where there is disagreement between the theoretical and the experimental values, one can explain the difference using the so-called *residual entropy* which is essentially the number of configurations that have nearly equal energies at very low temperatures (see Section 3.2).

Finally, note that the entire term under the logarithm must be a pure number. We must choose the units of each of the quantities in such a way that the entire expression becomes a pure number.

Conclusion

We started with the SMI defined on the locational and the velocity distribution of all the particles. Since the locations and the velocities of all the particles determine the microscopic state of the system, we can say that the SMI was built on the probability distribution of the microscopic states of the classical system.[21]

To obtain the entropy of the system from the SMI, we first calculated the distribution that maximizes the SMI. This is referred to as the equilibrium distribution; it is done by using the calculus of variations (see Ben-Naim, 2008). Having the equilibrium distribution, we can calculate the value of the maximum SMI. In order to obtain the entropy of the system, we simply have to multiply the maximum SMI by a constant. This constant consists of the Boltzmann constant and the conversion from the logarithm base 2 to the natural logarithm. Once we do this, we get the entropy function of an ideal gas (see Box 1.4).

This is the most amazing result. Starting with a quantity defined by Shannon in communication theory, a quantity which has nothing to do with physics, we get the entropy of a system defined in thermodynamics.

One final comment regarding the size of the system is in order. Remember that the SMI is defined for any distribution. Therefore, we can also define the SMI for a system of one, two, or three particles in a box. We can also proceed to define the corresponding entropy of such a system. However, if N is a small number, we shall find that the system will not obey the Second Law of Thermodynamics. In Section 1.5, we discuss the Second Law and relate it to entropy.

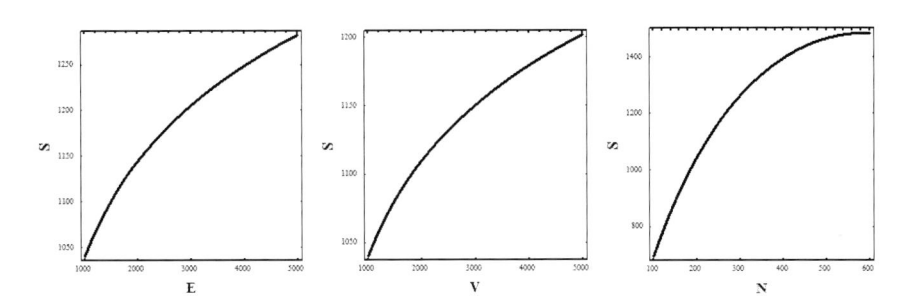

Fig. 1.16. The dependence of the entropy on E, V, and N.

We shall see that if we expect the system to approach an equilibrium state, and stay there "forever," then we need to take the SMI of a very large number of particles. Indeed, macroscopic systems contain a huge number of particles. This is the reason why we *never* observe large fluctuations from the equilibrium state.

In Figure 1.16 we show the entropy function obtained from the SMI, as the entropy, a monotonically increasing function of E, V, and N. Also, it has a negative *curvature*. The latter is an important characteristic property of the entropy. We shall discuss this property of the entropy in Chapter 2.

1.4.6 *Extension to Systems with Interacting Particles*

We have shown that the SMI defined on the probability distribution of the locations and velocities (or momenta) of all particles, leads, up to a multiplicative constant, to the entropy of an ideal gas.

It is possible in principle to extend the SMI approach to a system of interacting particles. It can be shown that whenever we add intermolecular interactions, the entropy of the system will *decrease*. This decrease can be cast into the form of mutual information. The result is that the entropy of any system can always be interpreted as an SMI. For details, see Ben-Naim (2008).

Numerically, it is very difficult to calculate the reduction in the SMI due to intermolecular interactions. This difficulty is not much different from the calculation of the entropy of a system of interacting particles using the standard methods of statistical mechanics.

1.5 The Entropy Formulation of the Second Law of Thermodynamics

In the previous sections of this chapter, we presented three different definitions of entropy. The only definition which also provides a clear, simple, intuitive, and proven interpretation of entropy is the one based on the SMI. This interpretation also expunges any traces of mystery associated with entropy. This is one reason why the definition based on the SMI is superior to the other two definitions.

Knowing what entropy is does not answer the question why it always increases under certain conditions (see below). In other words, having a well-defined and well-interpreted quantity called entropy does not provide an explanation of the Second Law.

This section is devoted to understanding the Second Law of Thermodynamics in isolated systems. We start with a simple example of a spontaneous expansion of a system of N particles from one compartment of volume V, to a larger volume $2V$. With this example we formulate the Second Law. We also show the relationship between the Second Law, and the SMI and the entropy. We then present a few, popular formulations of the Second Law and critically examine their validity. In Section 1.6 we shall discuss other formulations of the Second Law for systems which are not isolated.

It should be noted that the concept of entropy is defined in thermodynamics *only* for equilibrium states. Thus, when we speak about a spontaneous process, we do not mean a process from any state to any state, but from one equilibrium state to another equilibrium state. An equilibrium state is characterized by a small number of parameters such as temperature, volume, and pressure, far smaller than the number of parameters that are required to characterize the state of a system not in an equilibrium state.

An isolated system is characterized by a constant energy E, volume V, and number of particles of the species $N = (N_1, N_2, \ldots, N_c)$ where N_i is the number of particles of the species i. For such a system, the Second Law states that if we remove any constraint within the system in such a way that the system evolved from the initial constrained equilibrium state to the final unconstrained equilibrium state, the entropy will increase.

The reader should realize that this formulation of the Second Law is very different from the (meaningless) statement that entropy always increases! See Note 3 to the Preface.

For the expansion and for the mixing processes in Figure 1.2, removal of the constraint means the removal of the partition between the two compartments. For the heat transfer, it is the replacement of the insulating wall (athermal) by a heat conducting wall (diathermal). For a chemical reaction, it is the removal of an inhibitor or the addition of a catalyst. We now turn to discuss in more details the process of expansion of an ideal gas. Other processes are discussed in Chapter 3.

1.5.1 An Expansion of an Ideal Gas in an Isolated System

Consider the simplest spontaneous process. We start with an ideal gas confined to a volume V, remove a partition, and observe that the gas will expand and fill the entire new volume $2V$.

Ask any student who has learned thermodynamics: Why did the gas expand from V to $2V$, and why will it never go back to the original volume V? The answer you are most likely to hear is that the *cause* of that expansion is the tendency of the entropy to increase.

If you ask why the entropy tends to increase, the immediate answer would be: That is exactly what the Second Law states!

Does the tendency of the entropy to increase *drive* the spontaneous process, or does the spontaneous process drive the entropy upwards?

We will answer this question by examining the expansion process with a different number of particles. In all the following examples we shall assume that the particles do not interact with each other (or that interactions are negligible), and that in each case only the locational distribution of the particle is changed. Each particle is initially located within the boundaries of a volume V, and in the final state it is located in the larger volume $2V$.

There are essentially two questions associated with the Second Law that are oftentimes confused. The first one is why a system evolves spontaneously from one state to the other, and the second one is why entropy increases. The answers to these two questions are in principle different, yet they are related to each other.

The entropy formulation of the Second Law of Thermodynamics states that in any spontaneous process occurring in an isolated system the entropy increases. The Second Law does not state anything as to why the entropy increases, nor addresses the question as to why a spontaneous process occurs at all. Note again that by any "process," we mean a process from one equilibrium state to another equilibrium state.

As we shall see below, the only answer to the second question is probabilistic. Accepting the relative frequency interpretation of probability,[22] we can conclude that a system will be found more frequently in states which have higher probability. Specifically, for a thermodynamic system, we will see that the probability of finding the system in a state of equilibrium is almost one. Thus, we can say that a thermodynamic system at any initial state will always evolve toward a state of higher probability, and eventually reach a state we call *equilibrium state*, the probability of which is almost one. (For details, see Ben-Naim, 2008, 2012).

Again, we stress here that we talk about spontaneous processes in an isolated system, having a fixed energy, volume, and number of particles.

Answering the question of why the system evolves toward the state of equilibrium, leaves the question of why entropy increases unanswered. Nevertheless, because of the intimate relationship between the SMI of the system and the probability of the state of the system, the answers to the two questions are also related to each other. This is discussed next.

1.5.2 What Drives the System to an Equilibrium State?

Here, we shall answer the question of what drives the system to an equilibrium state by examining a simple example.

Consider a system of N non-interacting particles (ideal gas) in a volume $2V$ at constant energy E. We divide the system into two comparments L and R, each of volume V (see Figure 1.17). We define the microscopic state of the system when we are given $E, 2V, N$, and in addition we know which specific particles are in the right compartment (R), and which specific particles are in the left compartment (L). The macroscopic description of the same system is $(E, 2V, N; n)$ where n is the number of particles in the compartment L. Thus, in the microscopic description, we are given a specific configuration of the system as if the particles were labeled $1, 2, \ldots, N$. Here, by configuration we mean only which particles are in R and which are in L. (This is different from the more detailed configuration we discussed in Section 1.4). In the macroscopic description, we are given the information only on the number of particles in each compartment.

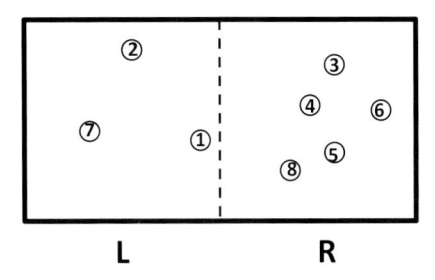

Fig. 1.17. A microscopic definition of the state of a system of eight particles. Here particles 1, 2, and 7 are in L, and 3, 4, 5, 6, and 8 are in R.

Clearly, if we know only that there are n particles in L, and $N-n$ particles in R, we have many specific configurations that are consistent with the requirement that there are n particles in L.[23]

We denote by $W(n)$ the number of specific configurations consistent with n particles in L. The first postulate of statistical mechanics states that all specific configurations of the system are equally probable. Clearly, the total number of specific configurations is 2^N, i.e. each particle can be in either one of the two compartments.[23] Using the classical definition of the probability, we can calculate the probability of finding n particles in L and $(N-n)$ particles in R. We denote this probability by $P_N(n)$.[24] It is easy to show that both $W(n)$ and $P_N(n)$ have a maximum as a function of n at the point $n^* = \frac{N}{2}$ (see below). The maximum value of the probability $P_N(n)$ (obtained at $n^* = \frac{N}{2}$) is denoted by $P_N(n^*)$.

Thus, for any given N, there exists an n, such that the number of configurations, $W(n)$, or of the probability, $P_N(n)$, is maximal. Therefore, if we prepare a system with any initial distribution of particles n, and $N-n$ in the two compartments, and let the system evolve, the system's state will change from a state of lower probability to a higher probability. As N increases, the value of the maximum number of configurations $W(n^*)$ *increases* with N. However, the value of the maximal probability $P_N(n^*)$ *decreases* with N.

To appreciate the significance of this fact, we will examine the "evolution" of systems with small numbers of particles. We shall see in what sense the spontaneous process of expansion proceeds in "one direction only," or is "irreversible." Later, we shall also follow the changes in the SMI in the process of expansion and finally, we shall calculate the entropy change for this process. For some simulations, see Ben-Naim (2008, 2010).

The case of two particles: $N = 2$

Suppose we have the total of $N = 2$ particles. In this case, we have the following possible configurations and the corresponding probabilities:

$$n = 0 \quad\quad n = 1 \quad\quad n = 2,$$
$$P_N(0) = \frac{1}{4}, \quad P_N(1) = \frac{1}{2}, \quad P_N(2) = \frac{1}{4}$$

This means that on the average, we can expect to find the configuration $n = 1$ (i.e. one particle in each compartment) about half of the time, but each of the configurations $n = 0$ and $n = 2$ only a quarter of the time (see Figure 1.18a). If we start with all the particles in the left compartment, we shall find that the system will "expand" from V to $2V$. However, once in a while the two particles will be found in one compartment.

The case of four particles: N = 4

For the case $N = 4$, we have the distribution as shown in Figure 1.18b. The maximal probability is $P_N(2) = \frac{6}{16} = 0.375$ which is smaller than $\frac{1}{2}$. In this case, the system will spend only $\frac{3}{8}$ of the time in the maximal state $n^* = 2$. Again, if we start with all particles in one compartment, the system will "expand" from V to $2V$, but once in a while we shall see all the particles in one compartment.

The case of ten particles: N = 10

For $N = 10$, the distribution is shown in Figure 1.18c. We calculate the maximum at $n^* = 5$ which is $P_{10}(n^* = 5) = 0.246$.

In all of the three examples we examined above, the system *expands* from V to $2V$. However, there is nothing in this process which can be termed "irreversible." In each case, the initial state will be visited once in a while.

Very large number of particles

Let us proceed with larger N. Figure 1.19 shows the probabilities $P_N(n)$ for larger number of particles. It is seen that the maximum value of $P_N(n)$ *decreases* as N *increases*.

It can be shown that the maximal probability decreases as $N^{-1/2}$. In practice, we know that when the system reaches the state of

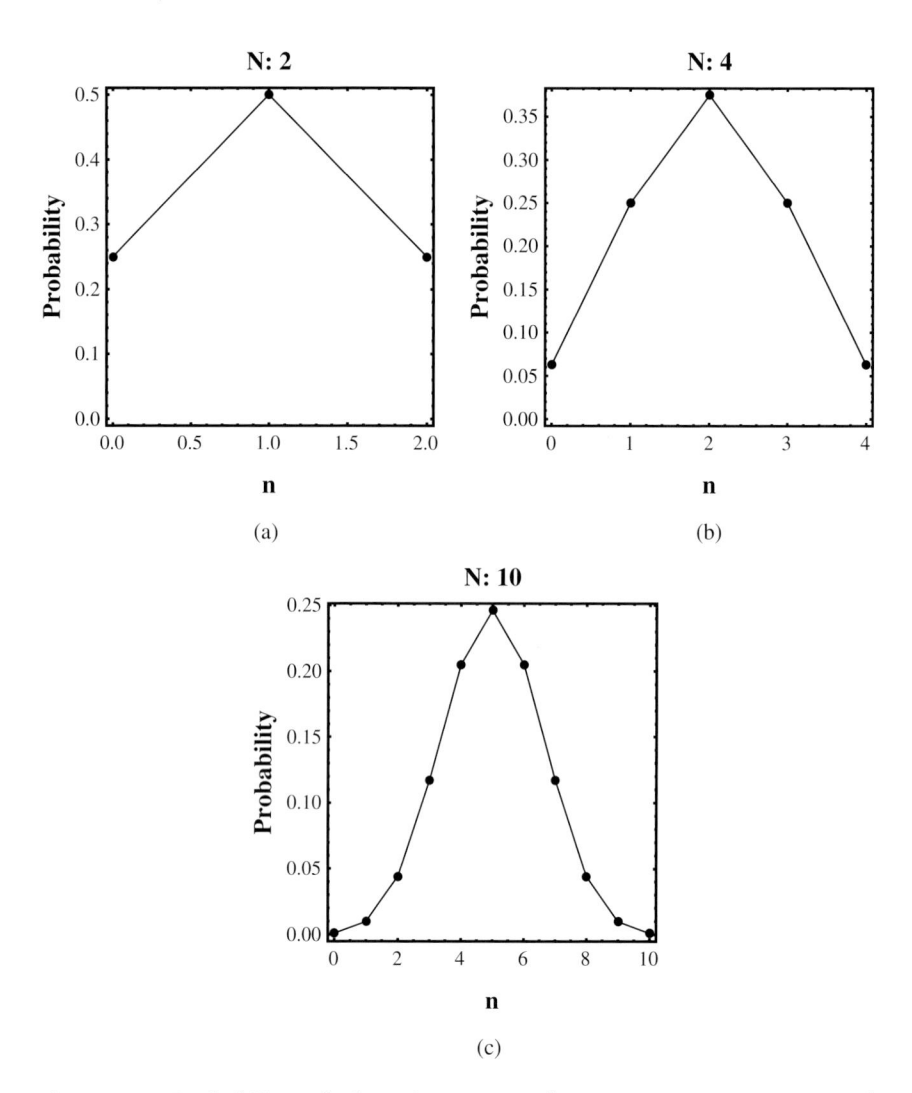

Fig. 1.18. Probability of observing n particles in one compartment and $N - n$ in the other for different numbers N.

equilibrium, it stays there *forever*. The reason is that the macroscopic state of equilibrium is not the same as the state for which $n^* = \frac{N}{2}$, but it is this state along with a small neighborhood of n^*, say $n^* - \varepsilon N \leq n \leq n^* + \varepsilon N$, where ε is a small number. For $N = 100$

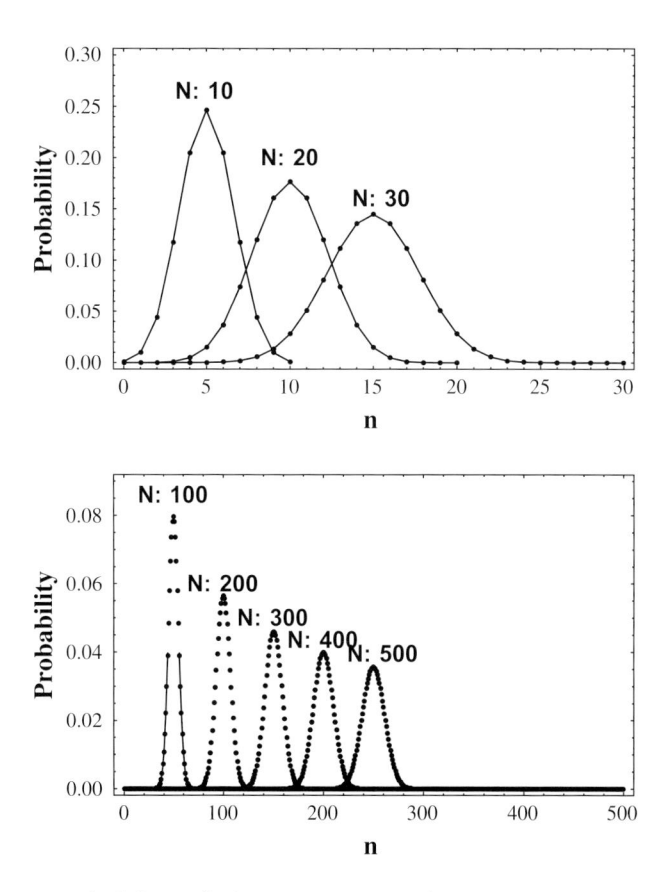

Fig. 1.19. Probability of observing n particles in one compartment and $N - n$ in the other, for different numbers N.

and $\varepsilon = 0.01$, the probability of finding n in the neighborhood of n^* is about 0.235. For $N = 10^{10}$ particles, we can allow deviations of 0.001% of N and the probability of staying in this neighborhood is nearly one. For more details, see Ben-Naim (2008, 2012).

In Figure 1.20, we show the probability of finding n between $n^* - \varepsilon N \leq n \leq n^* + \varepsilon N$ as a function of N, with $\varepsilon = 0.0001$. Plotting the probability $P_N(n^* - \varepsilon N \leq n \leq n^* + \varepsilon N)$ as a function of N shows that this probability tends to one as N increases. When N is on the order of 10^{23}, we can allow deviations of $\pm 0.00001\%$ of N, or even smaller, yet the probability of finding n at or near n^* will be almost

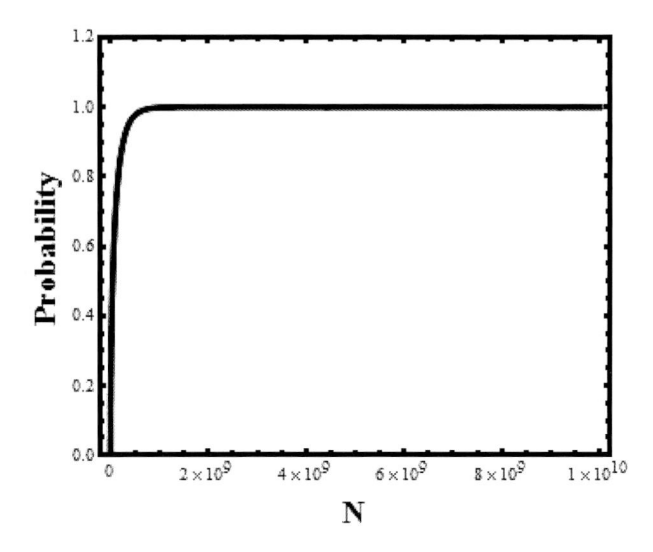

Fig. 1.20. Probability of finding n particles in the neighborhood of $n^* = N/2$, in one compartment as a function of N.

one. It is for this reason that when the system reaches n^* or near n^*, it will stay in the vicinity of n^* for most of the time. For N on the order of 10^{23}, "most of the time" practically means *always*.

The abovementioned specific example provides an explanation for the fact that the system will "always" evolve in one direction, and "always" stay at the equilibrium state once that state is reached. The tendency toward a state of larger probability is equivalent to the statement that events that are supposed to occur more frequently will occur more frequently. This is plain common sense. The fact that we do not observe deviations from either the monotonic climbing of n towards n^*, or staying close to n^*, is a result of our inability to detect small changes in n (or equivalently small changes in the SMI, see below). Note that in this section we did not say anything about the entropy changes. Before turning to calculate the entropy changes, we repeat the main conclusion of this section. For each N the probability of finding a distribution of particles $(n, N - n)$ in the two compartments L and R has a maximum of $n^* = \frac{N}{2}$. For a very large number of particles the probability of obtaining the exact value of $n^* = \frac{N}{2}$ is not very large. However, the probability of finding the system at a small vicinity of $n^* = \frac{N}{2}$ is almost one!

When we say that the system has reached an equilibrium state, we mean that we do not *see* any changes that occur in the system. In this example, we mean changes in the density of the particles in the entire system. In other experiments when there is heat exchange between two bodies, we characterize the equilibrium state as the one for which the temperature is uniform throughout the system and does not change with time.

At equilibrium the macroscopic density we measure at each point in the system is constant. In the particular system we discussed above the measurable density of the particles in the two compartments is $\rho^* \cong N/2V$. Note that fluctuations always occur. Small fluctuations occur very frequently, but they are so small that we cannot measure them. On the other hand, fluctuations that could have been measured are extremely infrequent, and practically we can say that they never occur. This conclusion is valid for very large N.

1.5.3 The Evolution of the SMI in the Expansion Process

Next, we will discuss the relationship between the probabilities of the macrostates and the formulation of the second law in terms of the entropy. We rewrite the essential quantities of the example discussed above in a slightly different way. Instead of n and $N-n$, we define the fractions $p = \frac{n}{N}, q = (1-p) = \frac{N-n}{N}$. p is the fraction of particles in the L compartment and $q = (1-p)$ is the fraction in the R compartment. We can also think of an ensemble of systems, all having the same macroscopic description in terms of $E, 2V, N$, but the ensemble is prepared in any arbitrary value of p (and hence, $1-p$). Clearly, the pair of numbers $(p, 1-p)$ is a probability distribution.

We now proceed to calculate the probability of this probability distribution. The number of specific configurations for which there are

68 | Entropy: The Truth, the Whole Truth and Nothing But the Truth

n particles in L is rewritten as $W(p,q) = \binom{N}{pN}$, and the corresponding probability is $\Pr(p,q) = \left(\frac{1}{2}\right)^N \binom{N}{pN}$.

This expression can be converted to obtain a relationship between the SMI of the system and the probabilities $\Pr(p,q)$.[25]

$$\Pr(p,q) = \left(\frac{1}{2}\right)^N \frac{2^{N \times SMI(p,q)}}{\sqrt{2\pi N p q}}$$

This is the relationship between the SMI defined on the distribution (p,q), and the probability $\Pr(p,q)$ defined on the same distribution. From the monotonic relationship between Pr and the SMI, it follows that whenever the SMI increases, Pr also increases, and at equilibrium both the SMI and Pr attain a maximal value. We have seen that the maximal value of SMI is related to the entropy of the system. Therefore, the answer to the question of why the entropy increases (in a spontaneous process in an isolated system), is the same as the answer to the question of why the state of the system evolves toward equilibrium; namely, it is because the probability Pr of the new equilibrium state is much larger. The connection between the questions of "what" entropy is, and "why" it changes in one direction, is shown symbolically in Figure 1.21.

Note carefully the two "levels" of probabilities. One is the probability distribution of a state described by (p,q). Pr is the probability of finding a state described by (p,q). To distinguish between the two probabilities, I refer to Pr as a *super probability*. Note also that the

$$WHY = 2^{N \times WHAT}$$

Fig. 1.21. Symbolic relationship between the "what" and the "why" questions.

answer to the question "Why does the system evolve toward equilibrium?" is provided by the probability Pr. Because of the monotonic relationship between Pr and the SMI, the answer to the question why the entropy increases is also probabilistic. It is easy to generalize this conclusion to the case of any number of compartments. See Ben-Naim (2008, 2012).

1.5.4 *Summary of Facts*

Before we turn to the general formulation of the Second Law, we summarize what we have found so far from the simple examples of expansion of N particles from volume V to $2V$.

For any N, right after removing the partition, we follow the evolution of the system with time. In all the examples, we observed that the particles which were initially confined to one compartment can access the larger volume $2V$. We can ask the following questions:

1. Why do the particles occupy the larger volume?
2. Does the number of particles in the left compartment change monotonically with time?
3. Does the number of particles in the left compartment reach an equilibrium state?
4. How fast does the system reach the equilibrium state?
5. How does the SMI of the system change with time?
6. How does the entropy change with time?

I urge the reader to try to answer these questions before continuing. Clearly, the answers to all these questions depend on N. Here are the answers to the questions:

1. The reason the particles will occupy the larger volume $2V$ rather than V is that the probability of the states where there are about $N/2$ in each compartment is larger than the probability of the state where all the particles are in one compartment. This is true for any N. When N is very small, there is a relatively large probability

that the particles will be found in one compartment. For these cases we cannot claim that the process is irreversible, in the sense that it will never go back to the initial state. For large N, even on the order 10^6, the probability to return to the initial state becomes so small that it is practically zero. However, there is always a finite probability that the system will visit the initial state. For N on the order of 10^{23}, the probability of visiting the initial state is so small (but still non-zero) that we can safely say that the system will never return to the initial state. "Never," in the sense of billions of ages of the universe.

2. The number of particles in L, n, does not change monotonically from N to N/2 (or from zero to N/2 if we start with all particles in the right compartments). Simulations show that for large values of N, the number n changes *nearly* monotonically toward N/2. The larger the N, the more monotonic is the change of n. (For simulated results, see www.ariehbennaim.com, my book *Entropy Demystified*, and simulated games). For N on the order of 10^6 or more, you will see nearly perfect, smooth, monotonic change in n.

3. The answer to this question depends on how one defines the equilibrium state of the system. If we define the equilibrium state when the value of n is equal to N/2, then for any n, when n reaches N/2, it will not stay there "forever." There will always be fluctuations about the value of $n^* = N/2$. However, one can define the equilibrium state as the state for which n is in the neighborhood of $n^* = N/2$. In such a definition, we will find that once n reaches this neighborhood, it will stay there for a longer time than in any other state. For N on the order of 10^6 or more, the system will stay in this neighborhood forever. Again, "forever" here means many ages of the universe.

4. The answer to this question depends on the temperature and on the size of the aperture we open between the two compartments. (In the experiment of Figure 1.2 we removed the partition between the two compartments. However, we could do the same experiment by opening a small window. In such an experiment, the speed of

reaching the equilibrium state would depend on the size of the aperture of the window.) In any case thermodynamics does not say anything about the speed of attaining equilibrium.

5. For each distribution of particles $(n, N - n)$ we can define a probability distribution $(p, 1 - p)$, and the corresponding SMI. As the system evolves from the initial state to the final state, n will change with time, hence also p will change with time, hence also the SMI will change with time. (For simulations, see Ben-Naim, 2010.)

   For small N, the SMI will start from zero (all particles being in one compartment) and will fluctuate between zero and N bits. When N is very large, say 10^6 or more, the value of SMI will change nearly monotonically from zero to N bits. There will always be some fluctuations in the value of SMI, but these fluctuations will be relatively smaller the larger N is. Once the system reaches the equilibrium state it will stay there forever. Note carefully that the SMI is defined here on the probability distribution $(p, 1 - p)$. For the initial distribution $(1, 0)$ the SMI is zero. The SMI defined on the distribution of locations and momenta is not zero. We shall further discuss this in Section 2.7.

6. The answer to this question is the simplest yet the most misconstrued one. It is the simplest because entropy is a *state* function, it is defined for a well-defined macroscopic (or thermodynamic) state of the system. For the expansion process, the macrostate of the system is defined initially by (E, V, N). The corresponding value of the entropy is $S(E, V, N)$. The final macrostate is characterized by $(E, 2V, N)$, and the corresponding value of the entropy is $S(E, 2V, N)$. In between the two macrostates (E, V, N) and $(E, 2V, N)$ the system's state is not well defined. A few intermediate states are shown in Figure 1.22. While E and N are the same as in the initial state, the "volume" during the expansion process of the gas is not well defined. It becomes well defined only when the system reaches an equilibrium state. Therefore, since the volume of the system is not well defined when the gas expands, the entropy is also not well defined. We can say that the entropy changes abruptly

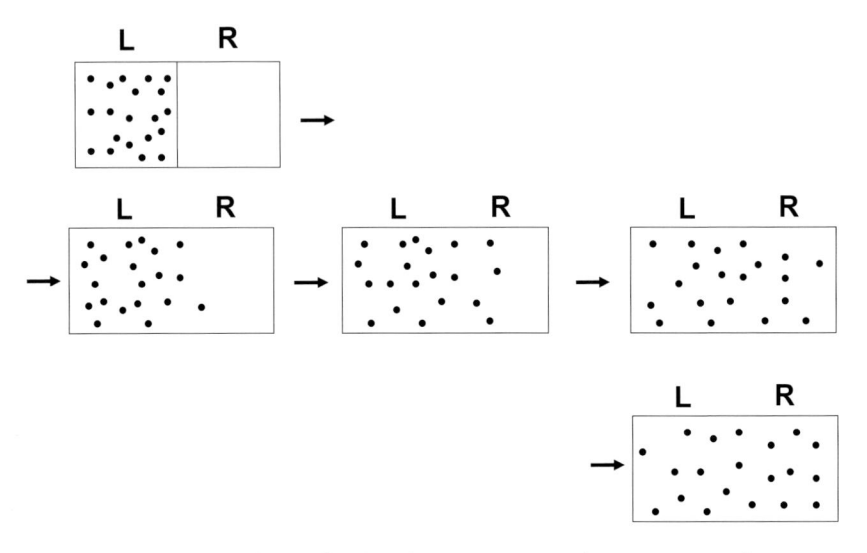

72 | Entropy: The Truth, the Whole Truth and Nothing But the Truth

Fig. 1.22. The initial, the final and some intermediate states in the expansion process.

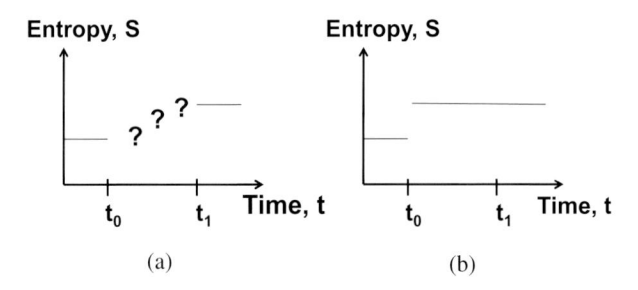

Fig. 1.23. Two views of the entropy change after removal of the partition.

from $S(E, V, N)$ to $S(E, 2V, N)$, and that this change occurred at the moment the system reaches a final equilibrium state. This is schematically shown in Figure 1.23a.

One can also adopt the point of view that when we remove the partition between the two compartments, the volume of the gas changes abruptly from V to $2V$; although the gas is initially still in one compartment, the total volume $2V$ is accessible to all particles. If we adopt this view, then at the moment we removed the partition, the volume changes from V to $2V$, and the corresponding change in entropy is $S(E, 2V, N) - S(E, V, N)$. This change occurs abruptly at the moment we remove the partition (see Figure 1.23b). Personally, I prefer the first point of view. Initially, it has the value $S(E, V, N)$ before the removal of the partition, and it reaches the value of $S(E, 2V, N)$ when the system reaches the new, final equilibrium state. In all the intermediate states the entropy is not defined. Note however, that the SMI is defined for any intermediate states between the initial and the final state. However, the entropy is the maximal value of the SMI (multiplied by the Boltzmann constant and change of the base of the logarithm), reached at the new equilibrium state.

It should be noted, however, that we could devise another expansion (referred to as *quasi-static*) process by moving gradually the partition between the two compartments. In this process the system proceeds through a series of equilibrium states, and therefore the entropy is well defined at each of the points along the path leading from (E, V, N) to $(E, 2V, N)$. In this process, the entropy of the gas will gradually change from $S(E, V, N)$ to $S(E, 2V, N)$ (see Figure 1.24). The length of time it takes to proceed from the initial to the final state depends on how fast or how slow we carry out the process.

Note that the sequences of states in the spontaneous process are different from those in the quasi-static process. In the latter, the states as well as the entropy of the gas are well defined along the entire path from the initial to the final equilibrium state, whereas in the

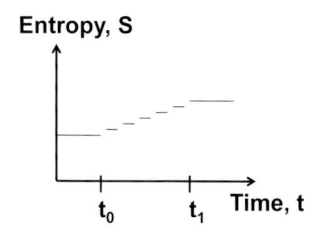

Fig. 1.24. The entropy change in a quasi-static process.

spontaneous expansion neither the states nor the entropy are defined along the path leading from the initial to the final state.

1.5.5 *Examples of Processes Associated with the Second Law*

There are as many formulations of the Second Law as there are interpretations of the entropy. We shall see that some formulations are very restricted, e.g. Clausius' formulation in terms of heat flow from a hot to a cold body. Others, are over-generalization formulations, e.g. again Clausius' statement that "the entropy of the universe always increases." In Section 1.6, I will present what I believe is the most general formulation of the Second Law.

Before I make any general statement of the Second Law, I want to emphasize that understanding entropy is not necessary to the understanding of the Second Law. One can understand what entropy is without understanding the Second Law. Similarly, one can understand the Second Law without understanding entropy.

In Table 1.1, a few processes commonly discussed in connection with the Second Law are listed.

All the processes listed in the table and many others are mentioned in popular science books discussing the Second Law.

The first three examples are spontaneous processes occurring in an isolated system (E, V, N constants). We can calculate the entropy change for these processes and find that it is positive. For these particular processes (and similar ones occurring in an isolated system) we can say that positive entropy change is a result of the occurrence of the process. One could also generalize and say that for any isolated system being in a constrained equilibrium state (e.g. a partition separating the two compartments), the entropy will increase as a result of the removal of the constraint. The last statement could be referred to as the entropy formulation of the Second Law. Note, however, that this formulation is restricted to isolated systems, and furthermore it does *not* provide

Table 1.1 Some Processes Mentioned in Connection with the Second Law of Thermodynamics

Process	The Associated Thermodynamic Function
1. Expansion of ideal gas in an isolated system	Entropy
2. Mixing two ideal gases in an isolated system	Entropy
3. Heat transfer from a hot to a cold body (The combined system is isolated)	Entropy
4. A spontaneous chemical reaction in a T, V, N system	Helmholtz energy
5. A spontaneous chemical reaction in a T, P, N system	Gibbs energy
6. Spontaneous mixing in a T, P, N system	Gibbs energy
7. Splattering of an egg	?
8. Melting of a candle	?
9. Cooling of soup	?
10. People aging	?
11. Remembering the past but not the future	?
12. Believing that we can affect the future but not the past	?
13. The existence of life	?
14. Artistic creativity	?
15. An untended child's room becomes messier	?

any explanation to the question why the entropy increases in such processes.

Process 4 in the table is a spontaneous process carried out in a system at fixed temperature, volume, and total number of atoms of each species [(T, V, N) constant]. An example could be the spontaneous formation of water (H_2O) from hydrogen (H_2) and oxygen (O_2) (see Chapter 3). In such a process, the entropy change can either increase or decrease. Instead, we can calculate the Helmholtz energy change ($\Delta A = \Delta E - T\Delta S$), and find that it is always negative in such a spontaneous process.

Similarly, processes 5 and 6 in the table are carried out at fixed temperature, pressure, and total number of particles [(T, P, N) constant]. Take for example a spontaneous folding of a protein in an aqueous solution (see Chapter 3). Again, in this process the entropy change can be either positive or negative or zero. However, the Gibbs energy change of such a process ($\Delta G = \Delta E + P\Delta V - T\Delta S$) will be negative.

Let us pause and see if we can find a common explanation for all these six processes. All these are spontaneous processes occurring in well-defined systems. However, we cannot apply the entropy formulation of the Second Law to all these processes.[26]

Instead, there is one common explanation for all these six processes. In all of these processes, the system proceeds from a state of lower to a state of higher probability. One should be careful to distinguish between the probability distribution of locations and velocities as we discussed in Section 1.4 and the probability of obtaining such a probability distribution. We have seen one relationship between the probability (Pr) of distribution (p, q), and the SMI of a system. For each system there is a probability distribution that maximizes the SMI. For such a distribution the value of the SMI (apart from a multiplicative constant) is the entropy of the system. Similarly, for process 4 there is a relationship between the probability, Pr, and the Helmholtz energy. For processes 5 and 6 there is a similar relationship between the probability, Pr, and the Gibbs energy.

Thus, we can conclude that in all the processes 1 to 6 in Table 1.1, the rationale for the occurrence of the spontaneous process is probabilistic. The system moves from a state having a relatively lower probability to a state of higher probability. The relationship between the maximum (Pr) and the thermodynamic function is different for different characterizations of the system (E, V, N), (T, V, N) or (T, P, N).

Let us briefly turn to processes 7 and 9. My guess is that all these processes can be explained probabilistically. The trouble is that these systems are too complicated; we cannot calculate the probabilities (Pr) of their states. We certainly cannot calculate the change in either entropy, Helmholtz energy, or Gibbs energy for these processes. Thus, although I believe that underlying these processes is a probabilistic reasoning, I cannot prove it.

Regarding processes 10–15, these are all discussed in the literature in connection with the Second Law. In my opinion these processes have nothing to do with the Second Law, and certainly not with entropy.

1.6 The Different Formulations of the Second Law of Thermodynamics

Historically, there were two formulations of the Second Law, even before the introduction of the concept of entropy.

The first is due to Clausius (1854).

> No process is possible for which the sole effect is that heat flows from a reservoir at a given temperature, to a higher temperature.

The second definition is due to Kelvin (1882).

> No device, operating in a cycle, can produce the sole effect of extraction of a quantity of heat from a heat reservoir and the performance of an equal quantity of work.

It is easy to show that these two formulations are equivalent, but note that neither one of these mentions entropy.

In this section we shall discuss in detail the different formulations of the Second Law for different systems (or different ensembles). We start with a short description of the various ensembles of systems. Then, we discuss the formulation of the Second Law for some of the most common ensembles. As we shall see, the entropy formulation is valid for the isolated system. For a system (or an ensemble of systems) characterized by the variables T, V, N, the Second Law is formulated in terms of the Helmholtz energy. Similarly, for a system (or an ensemble of systems) characterized by the variables T, P, N, the formulation of the Second Law is in terms of the Gibbs energy. Thus, each ensemble has a *different* characteristic function which attains an extremum (either a maximum or a minimum) at equilibrium. However, in all the ensembles the common "driving force" toward the new equilibrium state is probabilistic.

Note to the reader: The following outline is more mathematical and abstract. The reader can skip Section 1.6.1 and continue to Section 1.6.2.

1.6.1 *A Brief Survey of the Various Ensembles*

In both thermodynamics and statistical mechanics we start with an isolated system characterized by a fixed energy E, volume V, and number of particles N. If the system contains several components, then N is understood as the vector $N = (N_1, N_2, \ldots, N_c)$, where N_i is the number of particles of the type i. Starting with the variables (E, V, N), we can transform to new variables, say replacing the energy E by the temperature T, or replacing the volume V by the pressure P. In thermodynamics, these changes in the thermodynamic variables are achieved by the Legendre transformation [see Callen (1985)]. In statistical mechanics, it is achieved by using the Laplace transforms. Figure 1.25 shows the general scheme of the possible changes

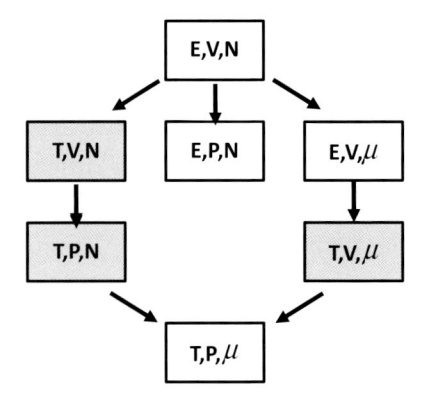

Fig. 1.25. The general scheme of changing variables. The more common variables are shown with grey background.

of variables. We shall briefly describe some of the most important ones below.

(a) The isolated system: The E, V, N ensemble

A truly isolated system does not exist. A real system is never perfectly isolated from its surroundings. Even if such a system existed, it would be of no interest to us. We could not make any measurements on it, nor observe its behavior. Any measurement or observation entails *interaction* with the system, and that is, by definition of the isolated system, impossible. Nevertheless, the isolated system is a convenient starting point for building up the theory of thermodynamics.

Let W be the number of quantum mechanical states of the isolated system characterized by the variables E, V, N. You do not need to know what these states are, or how one can calculate them. All you need to know is that a macroscopic system has a huge number of microscopic states (W), something of the order of N^N. If N is of the order of 10^{23}, which is huge, then N^N is unimaginably "huger" than N.

If you feel uncomfortable with the quantum mechanical state, think of a classical system. Each particle is characterized by its location and momentum (or velocity). A microstate of such a system is a list of all the locations and all the momenta of all the N particles; this is a huge vector $(R_1, R_2, \ldots, R_N, p_1, p_2, \ldots, p_N)$ where R_i and p_i are the locational and momentum vectors of particle i; (x_i, y_i, z_i) and (p_{ix}, p_{iy}, p_{iz}). Altogether, this vector has $6N$ components. Can you imagine how many states of such a system are consistent with the macroscopic characterization E, V, N? Do not try to count; there is an infinite number of states, presuming that this system behaves classically. The reason is that each of the variables R_i and p_i can change continuously.

In quantum mechanics the microstate of the system is described differently. Although it sounds more abstract than the description in terms of locations and momenta, the usage of the quantum mechanical states is easier in formulating the fundamental equation for the isolated system. Assuming that there are W states, and that these states are equally probable, the Boltzmann entropy is defined by $S_B = k_B \ln W$, where $k_B = 1.38 \times 10^{-23}$ J K^{-1} is called the *Boltzmann constant*. The fundamental distribution of the states in the isolated system is in fact, one of the postulates of statistical thermodynamics. It asserts that if there are W states, then the probability of being in state i is given by $\Pr(i) = p = \frac{1}{W}$. It is easy to show that Boltzmann's entropy as defined above can be written as an SMI over the distribution $\Pr(i) = \frac{1}{W}$, i.e. $S = -k_B \sum_{i=1}^{W} \Pr(i) \ln \Pr(i)$.

(b) The isothermal system: The T, V, N ensemble

Suppose we start with an ensemble of M isolated systems, each of which is characterized by the variables (E, V, N). We also assume that the systems in the ensemble do not interact with each other. In this ensemble we can interpret $\Pr(i)$ as the probability of finding the system in state i.

Now, suppose that the thermally insulated (or *athermal*) boundaries in the previous ensemble are replaced by thermally conducting (or *diathermal*) boundaries. The ensemble as a whole is still considered to be isolated. We allow the flow of heat between the individual systems in the ensemble, while still keeping the volume V and the number of particles in each system fixed.

What would happen?

Clearly, when the athermal walls are replaced by heat-conducting walls, heat will flow between the systems; and if we take a "snapshot" and measure the energy of each system in the ensemble, we shall find that the value of E is different for each system in the ensemble. When we "open" the boundaries between the systems to heat flow, the parameter E of each system in the ensemble is no longer fixed (but V and N remain unchanged). Instead there will be a distribution of energies. In other words, we can talk about the probability of finding a single system having an energy E, or equivalently the fractions of systems in the ensemble having some energy E (here we have assumed that the possible energies are discrete, say $E_0, E_1, E_2 \cdots$ and not continuous). We denote by $\Pr(E)$ the probability of finding a system with energy E, or the fraction of the systems in the ensemble having energy E. For macroscopic systems, the probability $\Pr(E)$ has a very sharp peak at the average value $\bar{E}$ which is equal to the fixed E of the initial isolated system.

The physical reason underlying this change in the thermodynamic variables is as follows. We know from experience that if we take two isolated systems and bring them to thermal equilibrium, heat will flow from the hot to the cold body. The energy of each system will no longer be fixed, but the two systems will be characterized by a fixed temperature. This is of course true for any number of systems. In our case we started with an ensemble of M systems, each characterized by the same energy E (as well as V and N). When we bring them to thermal contact, heat will be exchanged between the systems. The energy of each system will not be fixed, but will fluctuate (we can actually calculate the amount of these fluctuations). However, the temperature of the entire ensemble will be constant.

The formal transformation of variables from E to T (or equivalently from E, V, N to T, V, N) is achieved by using the discrete analog of the Laplace transform (or the Legendre transform in thermodynamics). We shall see below that the formulation of the Second Law in this case involves the Helmholtz energy $A(T, V, N)$ defined by $A = E - TS$. It is easy to show that the entropy of the system in the (T, V, N) ensemble can be written as $S = -k_B \sum_E \Pr(E) \ln[\Pr(E)]$.

(c) The isothermal isobaric system: The T, P, N ensemble

We now proceed to the next change of a thermodynamic variable. We start with an ensemble of systems characterized by (T, V, N). Each system has a fixed volume V. Remember, in the previous transformation we started with systems having fixed value of E, and we "opened" the boundaries of the systems so that heat can flow from one system to another. Similarly, we now "open" the boundaries of the system so that "volume can flow" from one system to another. What that means is that we replace the solid, rigid boundaries by some flexible boundaries. The total volume of the ensemble is again fixed, but the volume of each system in the ensemble can fluctuate. Therefore, we can talk about the distribution of volumes. For simplicity assume that the volume of the system can attain only discrete values, say $V_0, V_1, V_2, \ldots$. We denote by $Pr(V)$ the probability of finding a single system in the ensemble having a specific volume V. This function also has a sharp peak at the average value $\bar{V}$.

The formulation of the Second Law for a system characterized by T, P, N involves the Gibbs energy $G(T, P, N)$. Here, G is the Gibbs energy, expressed as a function of the variables T, P, N. It is defined by $G = E + PV - TS$. One can also show that the probabilities $Pr(V)$ determine the entropy of a system characterized by the variables (T, P, N), i.e. $S = -k_B \sum_V Pr(V) \ln[Pr(V)]$.

(d) The open system: The T, V, μ ensemble

We shall briefly discuss here the case of an open system. This case is important in statistical thermodynamics. It is less often used in connection with the Second Law.

Previously, we transformed from the variable T, V, N to T, P, N (i.e. from V to P). Here we start again with the T, V, N ensemble, and open the systems to a flow of particles. Clearly, once we replace the impermeable boundaries by permeable ones, particles can flow from one system to another. The total number of particles in the entire ensemble is still fixed, but in each system the number of particles will fluctuate. However, the chemical potential μ of each system in the ensemble is fixed.[27] In this ensemble we may ask for the probability of finding a system with exactly N particles. We denote by $Pr(N)$ the probability of finding the system with exactly N particles in the open ensemble. One can show that the entropy of such a system is given by $S = -k_B \sum \Pr(N) \ln \Pr(N)$.

Thus, we see that in each ensemble the entropy has the same form as the SMI defined on a probability distribution relevant to that ensemble. We emphasize that whenever we interpret entropy in terms of an SMI, we must use the probability distribution which maximizes the SMI for that system. Or equivalently, the entropy of the system is equal (except for a multiplicative constant) to the SMI of that relevant system at equilibrium.

1.6.2 The Formulation of the Second Law for Isolated Systems

If you ask anyone who has learned thermodynamics what the Second Law states, it is most likely that you will get this answer: "Entropy always increases until it reaches a maximum at equilibrium."

A more cautious person will say, "In a spontaneous process in an isolated system, the entropy increases and reaches a maximum at equilibrium." Both answers are incorrect. Entropy in itself does not increase or decrease. Entropy of an isolated system is well defined. The entropy of an isolated system does not change and does not reach a maximum! The entropy of an isolated system is the maximal value of the SMI (multiplied by a constant). The correct answer is that when we remove a constraint in a constrained equilibrium state of an isolated system, the entropy increases. This answer is correct, but it is not the most general formulation of the Second Law. It only applies to isolated systems. In daily life, you are more likely to encounter systems which are not isolated. The more common ones are either systems at constant temperature, pressure, or chemical potential.

Besides, saying that the system reaches a maximum entropy leaves the question "maximum, with respect to what" open. This "open question" is a very slippery pitfall in which many authors have fallen into. The most common view is that entropy is a function of time. And indeed, many authors who write on the Second Law identify the change of entropy with the so-called Arrow of Time. Some even claim that entropy is time itself![28]

Here we shall clarify the question regarding the parameters with respect to which the entropy attains a maximum. The formulation of the Second Law for an isolated system is as follows: We start with a *constrained* equilibrium state, e.g. one of the states on the left-hand side of Figure 1.26. When we remove the constraint, the entropy will reach a new equilibrium state having higher entropy.

Figure 1.26a shows a few cases of constrained equilibrium states. The system as a whole is isolated. Clearly, there are infinite numbers of constrained equilibrium states. The general formulation of the Second Law is that whenever we remove one or more constraints, the entropy can only increase. Callen (1985) formulated the Second Law for isolated systems in terms of the maximum entropy over "the manifold of constrained equilibrium states." Instead of such an abstract formulation, let us be more specific with what we mean by "constrained equilibrium states." Figure 1.26a shows three isolated systems. All

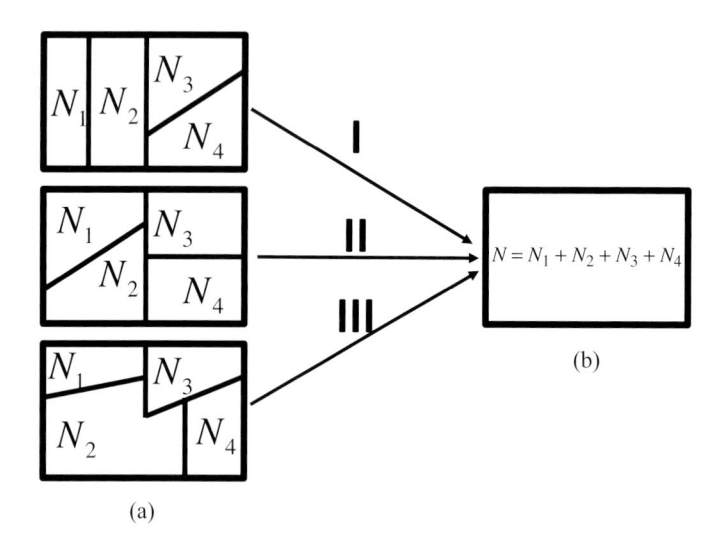

Fig. 1.26. Four different constrained equilibrium systems, all having the same total E, V, and N, and, on the right side, the unconstrained equilibrium state obtained after the removal of the constraints.

have the same total energy E, volume V, and number of particles N. In Figure 1.26b, we remove the constraints and we have a new equilibrium state having the same total E, V, and N, as in the constrained equilibrium states shown in Figure 1.26a. For this process, the Second Law states that the entropy of the system in Figure 1.26b has a maximum over all possible constrained equilibrium states, such as the ones shown in Figure 1.26a.

Note also the entropy changes of the three processes I, II, and III are different. For instance, for process I the entropy change is

$$\Delta S = S(E, V, N) - S_1(E_1, V_1, N_1) - S_2(E_2, V_2, N_2) \\ - S_3(E_3, V_3, N_3) - S_4(E_4, V_4, N_4)$$

Note that $E = E_1 + E_2 + E_3 + E_4$, $V = V_1 + V_2 + V_3 + V_4$ and $N = N_1 + N_2 + N_3 + N_4$ are unchanged in processes I, II, and III. The partitioning of E, V, and N into "components" is different for each case. Therefore, the entropy change from the initial constrained equilibrium states, to the final unconstrained equilibrium state, is different. We next turn to a more detailed discussion of three processes; in the first, only the locational distribution changes; in the second, only the velocity distribution changes; and in the third, both the locational and the velocity distribution change. In all of the following examples we remove a constraint and observe the change from one equilibrium state to another equilibrium state.

(i) **Change in locational distribution only**

In Figure 1.27a, we prepared a system with four subsystems having volumes V_1, V_2, V_3, V_4, and number of particles N_1, N_2, N_3, N_4. This system will be referred to as a *constrained* equilibrium state. If we remove one of the partitions, we shall proceed to a new constrained equilibrium state. The entropy of the new state will be higher. If we remove the second partition, we get a new constrained equilibrium state, and when we remove the third partition, we will obtain an *unconstrained* equilibrium state. Clearly, in each step when we

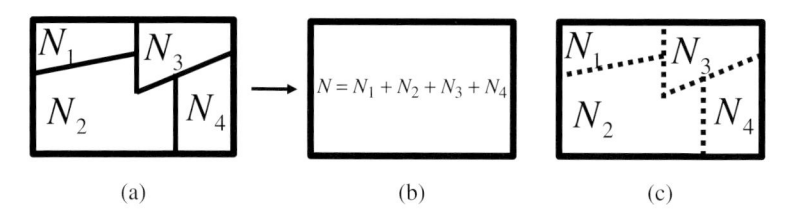

86 | Entropy: The Truth, the Whole Truth and Nothing But the Truth

Fig. 1.27. (a) A constrained equilibrium system, (b) the unconstrained equilibrium of the same system as in (a), and (c) the same system as in (b) but with a distribution of particles as in (a).

remove a partition, we get a new equilibrium state which is unconstrained relative to the previous state. In each of these steps the entropy will either increase or will stay constant (for instance, if the two densities ρ_1 and ρ_2 are equal, then removing the first partition will cause no change in entropy).

We shall refer to the partitioning of N into $N_1 + N_2 + \cdots + N_n$ as a distribution of particles in the n compartments $1, 2, \ldots, n$. The densities are defined by $\rho_i = N_i/V_i$, where V_i is the volume of the ith compartment. The mole fractions are defined by $x_i = N_i/N$. The vector $(x_i, x_2, \ldots x_n)$ is also a probability distribution.

Clearly, we have infinite possible constrained equilibrium states. Each of these are characterized by a probability distribution $(x_1, x_2, \ldots, x_n)$. The Second Law states that if there are no constraints (of the kind shown in Figure 1.27a), then the entropy of the system denoted $S(E, V, N)$ is maximal with respect to all possible constrained equilibrium states characterized by distributions $(x_1, x_2, \ldots, x_n)$.

As we have discussed in Section 1.5, the reason, or the "driving force" for the system to move from the constrained equilibrium state to the unconstrained equilibrium state is that the latter state is an overwhelmingly more probable state. Specifically, if the constrained equilibrium state is characterized by the distribution $(x_1, x_2, \ldots, x_n)$, then one can assign probability value to this distribution which we denote by $\Pr(x_1, x_2, \ldots, x_n)$. Also, to this distribution we can define the entropy $S(x_1, x_2, \ldots, x_n)$.

The Various Definitions of Entropy | 87

> $\Pr(x_1, x_2, \ldots, x_n) = C\exp[S(x_1, x_2, \ldots, x_n)/k_B].$
>
> $\Pr(x_1, x_2) = C\exp[S(x_1, x_2)/k_B].$
>
> $\Pr(\mathbf{x}) = C\exp[-G(\mathbf{x})/k_B T]$

Box 1.5

It should be emphasized that the probability Pr and the entropy S are related to each other by the equation in Box 1.5, $\Pr(x_1, x_2, \ldots, x_n) = C\exp[S(x_1, x_2, \ldots, x_n)/k_B]$. (C is a normalization constant). The two quantities Pr and S are *defined* for two different systems. The entropy is defined for the constrained equilibrium state when we prepare a system with partitions, with the specific distribution $(x_1, x_2, \ldots, x_n)$ which is maintained by the presence of the partitions (Figure 1.27a). On the other hand, Pr is defined for a system at equilibrium without partitions, for which we can ask what the probability is, that a distribution $(x_1, x_2, \ldots, x_n)$ will be found in the system without the partitions.

Figure 1.27b shows the unconstrained (i.e. after the removal of all the partitions in Figure 1.27a) equilibrium state. For this state, the entropy is $S(E, V, N)$, and this is the maximal entropy over all constrained equilibrium states of the kind shown in Figure 1.27a.

In system (b), we might observe fluctuations in the distribution of particles. One such fluctuation is shown in Fig. 1.27c. In this event the temporary distribution is $(x_1, x_2, \ldots, x_n)$. For the system (b) at equilibrium, we ask: "What is the probability of observing the distribution $(x_1, \ldots, x_n)$ as depicted in (c)?" However, note that (b) and (c) are the same system; neither has any internal constraints (no partitions). We ask for the probability of occurrence of $(x_1, \ldots, x_n)$ as shown in (c), but without the partitions. The dashed partitions in (c) are not real partitions; they only serve to clarify the distribution for which we define the probability $\Pr(x_1,\ldots,x_n)$. This probability, defined in system (b), is related to the entropy of the system in Figure 1.27a by a relationship of the type shown in Box 1.5.

For each distribution $(x_1,\ldots,x_n)$ one can define the entropy provided that it is an equilibrium state, i.e. with partitions which imposes the constrained equilibrium state. On the other hand, the entropy of the unconstrained system is the maximum over all possible constrained equilibrium states characterized by $(x_1,\ldots,x_n)$.

Note carefully that if we start with the constrained equilibrium state in Figure 1.27a, and we remove all the partitions, at this moment the system is not in an equilibrium state. The entropy of this system is not defined. However, one can define the SMI for this system (or for any other arbitrary distribution). After removing the partitions, the system will evolve to a new equilibrium state. At each moment of this evolution one can define the corresponding SMI. It is only when the system reaches its new equilibrium state that the SMI reaches its maximal value. This maximal value is proportional to the entropy of the system.

Pause and ponder

The example we have discussed above is, of course, not the most general case of a process for which the entropy formulation of the Second Law applies. To make sure that you fully understand this example, I urge you to examine a simpler case shown in Figure 1.28. Here, instead of four compartments, we only have two. We prepare a system with distribution (x_1, x_2) $(x_1 = \frac{N_1}{N}$ and $x_2 = \frac{N_2}{N}$, $x_1 + x_2 = 1$, $N_1 + N_2 = N)$. We remove the partition (the heavy line in Figure 1.28a), and we get a new equilibrium state shown in Figure 1.28b.

Now answer the following questions:

1. What is the entropy of the system in Figure 1.28a?
2. What is the entropy of the system in Figure 1.28b?

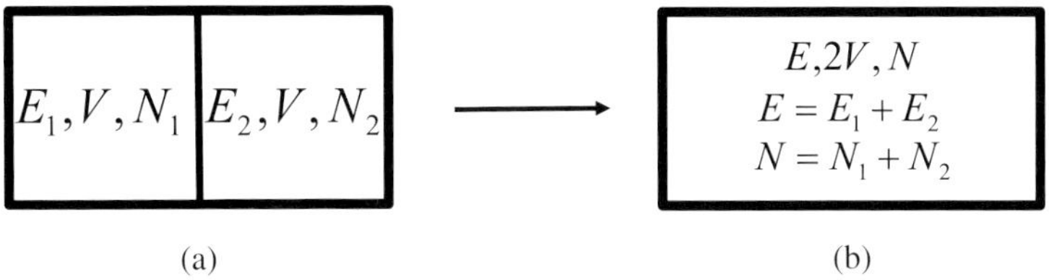

The Various Definitions of Entropy | 89

E_1, V, N_1 | E_2, V, N_2 ⟶ $E, 2V, N$
$E = E_1 + E_2$
$N = N_1 + N_2$

(a)   (b)

Fig. 1.28. (a) A constrained equilibrium system, and (b) the unconstrained equilibrium of the same system as in (a).

3. What is the probability of finding the system in Figure 1.28a with an arbitrary distribution of particles (x_1, x_2)?
4. What is the probability of finding the system in Figure 1.28b with an arbitrary distribution of particles (x_1, x_2)?

In this particular example we assume that in the initial state (with the partitions), the volumes V_1 and V_2 are the same and equal to V. The total energy of the system is E and the total number of particles is fixed: $N = N_1 + N_2$. For an ideal gas you can actually calculate all the relevant entropies and the probabilities. Here I ask you to give only qualitative answers to make sure you understand the difference between the systems in 1.28a and 1.28b. After writing down your answers compare them with my answers in Note 29.

(ii) Change in velocity distribution only

The second example involves constraints with respect to velocity distributions. Consider an isolated system characterized by $(E, 2V, N)$. It is initially divided in such a way that there are two subsystems separated by an athermal partition. Initially, we referred to the state in Figure 1.29a as a constrained equilibrium state. It is constrained in the same sense as before. At the initial state, the entropy of the system is simply the sum of the entropies of the two subsystems.

If we remove the partition in Figure 1.29a, the entropy will increase (unless $T_1 = T_2$). We know that heat will flow from the subsystem with higher temperature to the one with lower temperature, and the entropy of the system will increase. After we remove the

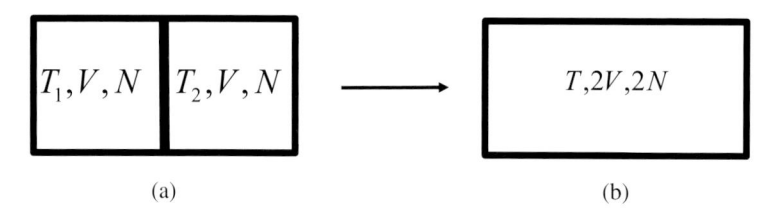

Fig. 1.29. (a) A constrained equilibrium system, with different temperatures, and (b) the unconstrained equilibrium after removing the athermal partition. The final temperature is T.

partition (constraint) the system will reach a new equilibrium state having a larger entropy, compared to the initial constrained equilibrium state. We also know that in the final state the temperature of the entire system will be constant, T.

Note that since we assumed that the initial densities in the two subsystems in Figure 1.29 are the same, the change of entropy in the process of removing the partition will be due to the heat transfer between the two compartments.

On a molecular level the entropy change may be ascribed to the change in the distribution of velocities of all the particles in the system. To see this, suppose for simplicity that we have N atoms of argon in each compartment. In this case, in each subsystem we have different velocity distributions, as shown in Figure 1.30a. We can also calculate the velocity distribution of all the atoms in the system which in this case is simply the average initial distribution. After the removal of the constraint (the partition) there will be one distribution of velocities which is determined by the final temperature T, as shown in Figure 1.30b. One can prove that the SMI of the final velocity distribution will be larger than that of the initial average distribution of velocities. The same is true for the entropy, which is defined both in the initial state and in the final equilibrium state. For more details on this process, see Ben-Naim (2008, 2012).

We can now generalize for any initial constrained equilibrium distribution of velocities, i.e. we can have any number of subsystems,

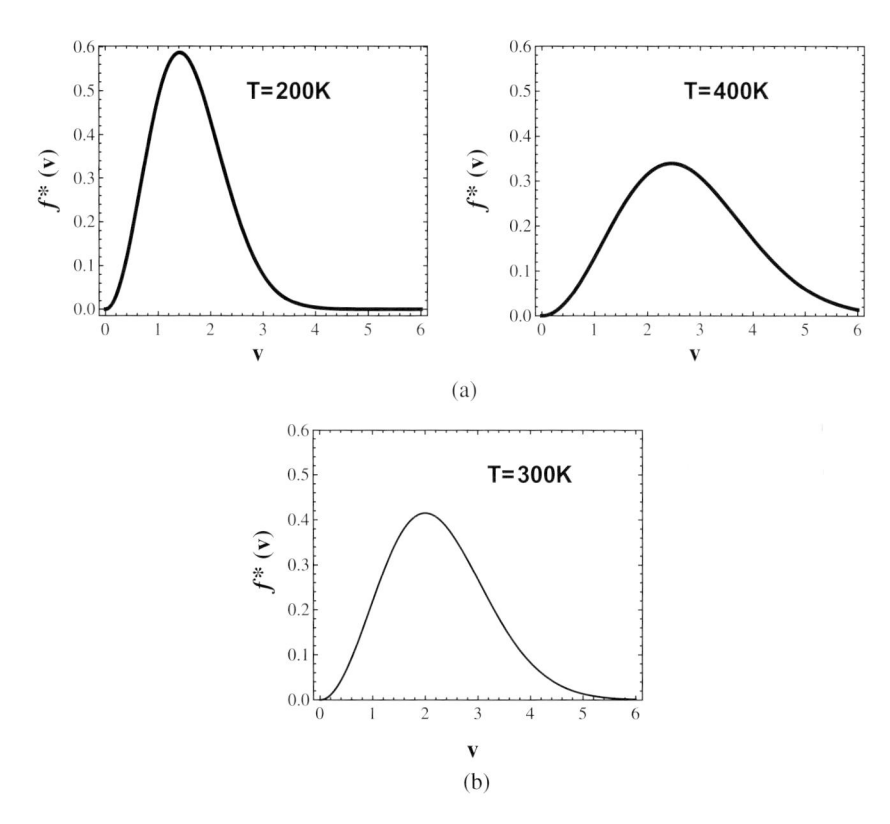

[Graphs showing f*(v) vs v at T=200K and T=400K labeled (a), and T=300K labeled (b)]

Fig. 1.30. The absolute velocity (speed) distribution of the two systems at the initial state (a) and in the final state (b).

each at different temperatures. The entropy of the unconstrained system is maximum with respect to all possible initial constrained equilibriums.

(iii) Removal of the constraints on both location and velocities of the particles

In the previous examples we treated separately the case of the spatial redistribution, and the case of velocity distribution, of the particles. We now discuss the more general case. Suppose we start again with a system with (E, V, N). We construct any constrained equilibrium state by imposing partitions between any number of subsystems. Each subsystem has a different density and different temperature. All we require is that the total energy E, volume V, and N is the same as in the unconstrained equilibrium state.

Note carefully that the SMI is defined for any distribution. In Section 1.4 we defined the SMI on a distribution of locations and velocities of a macroscopic system. We did *not* use the Second Law. We did *not* assume that the SMI has a maximum. Instead, we *found* that there is a distribution for which the SMI is maximal. We also found that the distribution which maximizes the SMI is the distribution which has the maximum probability. Denoting by x^* the distribution which maximizes the SMI, we found that the maximum value of the SMI, i.e. $H[x^*]$ is equal (up to a multiplicative constant) to the entropy of the ideal gas at equilibrium. Thus, the entropy of the ideal gas is defined as the maximum value of the SMI over all possible distributions x (of locations and velocities).

It is important to realize that so far, we have defined the entropy without any reference to the Second Law. We only used the property of the SMI which attains a maximum value for a specific distribution x^*. The entropy was defined at the specific state of equilibrium. When we want to formulate the Second Law in terms of the entropy, we *cannot* say that the entropy is maximal over *all* possible distributions x.

The reason is that the entropy is not defined for *any* arbitrary distribution x. Instead, we can define the entropy on a subset of distributions x. Each distribution in this subset corresponds to a constrained equilibrium state. The constraint is essential in maintaining an equilibrium state having the distribution x. Now we can formulate the Second Law by stating that the entropy of the system, characterized by E, V, N, is maximal over all constrained equilibrium states, or equivalently on the subset of distributions x, for which constrained equilibrium states exist.

It is sometimes said that the system evolves to the new equilibrium state *because of* the tendency of entropy to increase. This is incorrect. Entropy has no tendency to increase or decrease. A better formulation is that once we release the constraints, the system will evolve toward a state which has a higher probability. As a result of this change (toward a state of higher probability), the entropy increases.

As we shall see below, the probabilistic formulation of the Second Law is applicable to any well-defined thermodynamic system, whereas the entropy formulation applies only to isolated systems. Therefore, the probabilistic formulation of the Second Law is far more general than the entropy formulation.

1.6.3 The (T, V, N) Formulation of the Second Law

We shall briefly formulate the Second Law for systems characterized by the variables T, V, N. Here, the Second Law is formulated in terms of the Helmholtz energy A defined by $A = E - TS$. Having a system characterized by T, V, N, the Helmholtz energy of the system will have a minimum compared with all possible constrained equilibrium states. In other words, starting with any constrained equilibrium state, then removing the constraint (keeping T, V, N unchanged), the Helmholtz energy of the system will decrease.

Note that in such a process the entropy of the system can either increase or decrease. The Second Law does not say anything about the sign of the entropy change, only that the Helmholtz energy must decrease.

As in the case of the isolated system, the rationale for the change of the state of the system from the initial to the final equilibrium state is again, probabilistic. The system will evolve from a state having a relatively lower probability to a state of higher probability. We also note that the probability of the state is related to the Helmholtz energy by a similar relationship as that between the probability and the entropy in an isolated system. We shall elaborate on the details of this relationship in the next, more common, case of a system at constants T, P, N.

1.6.4 The (T, P, N) Formulation of the Second Law

Most of the experiments carried out in the laboratories in chemistry and biochemistry are at a fixed temperature and under atmospheric pressure. In such systems the formulation of the Second Law is exactly as in the previous case, except that we use the Gibbs energy instead of the Helmholtz energy. The Gibbs energy is defined by $G = A + PV = E - TS + PV$.

Because of its central importance in chemistry and biochemistry, we will discuss in Section 3.3.3 the process of protein folding. Here, we discuss a simpler system for which we apply the Second Law.

Our system is a solvent, say water, and a solute which can be in many configurations. Suppose we have N_W water molecules and N_P solute molecules. We also assume for simplicity that $N_P \ll N_W$, i.e. the solute is very diluted in the water, and we can neglect the solute-solute interactions. The system is maintained at a fixed temperature T and pressure P, as well as N_P and N_W. In general, the solute, say a protein, can have many conformations. Each conformation is described by the set of internal rotational angles.

Here we treat a simple solute having only one internal rotational degree of freedom, as shown in Figure 1.31. We also assume that there are n possible conformations.

We construct a constrained equilibrium state by fixing the conformation of the solute molecule. We can imagine that we can prepare all the solute molecules at a fixed conformation, say ϕ_0, and that we have some inhibitor that prevents the solute from changing

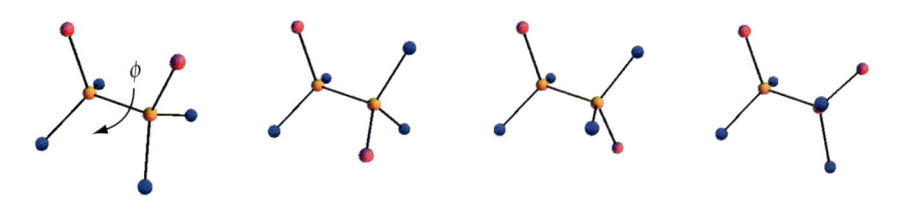

Fig. 1.31. A finite number of conformations of substituted ethane.

its conformation. Here, the inhibitor is the analogue of the partition between two compartments which we use to prevent the flow of energy, volume, or number of particles from one compartment to another. Similarly, the inhibitor prevents the "flow," or the conversion from the initial conformation ϕ_0 to any other conformation ϕ.

Now suppose that we prepare a system of N_W water molecules and N_P solute molecules in such a way that there are N_1 solute molecules at conformation ϕ_1, N_2 at conformation ϕ_2, and so on. We assume for simplicity that we have only n possible conformations $\phi_1, \phi_2, \ldots, \phi_n$. The distribution of species $(N_1, N_2, \ldots, N_n)$ may be translated into a probability distribution $(x_1, x_2, \ldots, x_n)$ where $x_i = N_i/N_p$.

In the presence of an inhibitor, the initial distribution $x^{in} = (x_1^{in}, x_2^{in}, \ldots, x_n^{in})$ will be maintained, much as the partitions in the case of Figure 1.27. For each of this initial distribution we can define the Gibbs energy of the system of the constrained equilibrium state $G(T, P, N_W, N_P; x_1^{in}, x_2^{in}, \ldots, x_n^{in})$.

When we remove the inhibitor (the analogue of removing a partition between two compartments), the system will evolve to a new equilibrium state. At this new equilibrium state the Gibbs energy of the system is lower than the Gibbs energy of the initial constrained equilibrium state. The new equilibrium distribution has changed from the initial distribution x^{in} to the final distribution x^{eq}, which is the distribution at the final equilibrium state.

Thus, the process that occurs in the system is $x^{in} \rightarrow x^{eq}$, and corresponding to this process, the Gibbs energy change is negative. It is easy to prove that the equilibrium distribution x^{eq} of conformations is the distribution that minimizes the Gibbs energy. [For a proof, see Ben-Naim (2016b)].

We can now formulate the Second Law for a system characterized by T, P, N (here T, P, N_W, N_P). The Gibbs energy of the system has a minimum over all possible constrained equilibrium states (in the example above, any fixed initial distribution in the presence of an inhibitor).

As we noted in connection with the entropy formulations, most people who use the Second Law for a T, P, N system would explain that the evolution of the system occurs because of the lowering of the Gibbs energy. A better way of rationalizing the direction of the evolution of the system is in terms of probabilities. The system moves from one equilibrium state to another because the new equilibrium state has a higher probability. The probability of the state (or the probability of a given distribution of conformations x) is related to the Gibbs energy by $\Pr(x) = C \exp[-G(x)/k_B T]$ (see Box 1.5). Thus, the distribution which minimizes the Gibbs energy also maximizes the probability.

Note carefully that the Gibbs energy is not a function of time. It is defined only for equilibrium states.

Thus, for the three cases discussed above, there exists a relationship between the probability of observing a specific distribution and a thermodynamic function (S, A, or G depending on the conditions of the experiment). In all of these cases the system will move from one equilibrium state to another equilibrium state. As a result of this change of state, there is a corresponding change in entropy, Helmholtz, or Gibbs energy according to whether the system is characterized by (E, V, N), (T, V, N) or (T, P, N), respectively. The common "driving force" underlying all these changes is probability.

1.7 A Few Misdefinitions of Entropy and the Second Law

The literature is crawling with inaccurate definitions of entropy. Sometimes the border between an interpretation and a definition is blurred. For example, Lambert writes[30]

> Entropy change is the measure of how more widely a specific quantity of molecular energy is dispersed in a process, whether isothermal gas expansion, gas or liquid mixing, reversible heating and phase change or chemical reactions.

Concisely, the Second Law is "Energy of all types changes from being localized to becoming spread out, dispersed in space if that energy is not constrained from doing so."

The first part of the quotation seems to be an interpretation of entropy. We shall discuss this interpretation in Chapter 2. The second part seems to define the Second Law in terms of spreading and dispersion of energy. Of course, such a description is incorrect and cannot serve as a definition of the Second Law.

In a more recent book by Kafri and Kafri (2013), we find a new and "original' definition of entropy as

$$S \geq \frac{Q}{T}$$

adding that the equality holds when in a state of equilibrium.[31] Regarding the inequality $S \geq \frac{Q}{T}$, the authors claim that it holds (1) in irreversible processes, and (2) for a system not at equilibrium.

Both of these claims are unfounded. First, because the entropy is not defined for a *process* (either reversible or irreversible), but for a *state* of a system at equilibrium, and second, for non-equilibrium states, neither Q nor S are defined. The quantity Q is not defined for a system at equilibrium. It is a quantity of heat which is *exchanged* (into or out from) with the system. If one defines the entropy as Q/T at equilibrium, then it makes Q a *state function*, which it is not![31] I wrote to the author about this absurd result and he wrote back saying that Q is indeed a quantity of heat which either flows into, or out from the system. Therefore, he said, "entropy has the same property as Q," which is an even greater absurdity than what I initially thought of about this definition.

Another misdefinition is found in Mayhew (2015a, 2015b). On the first page of his book, one finds the "definition" of entropy in the equation $TS = \varepsilon + PV$, where T is the absolute temperature, ε "signifies the internal energy of the system," P is the pressure and V is the volume.

To the best of my knowledge, the combination of $\varepsilon + PV$ is the definition of *enthalpy*, not of entropy.

In many popular science books one finds a "definition" of the Second Law as the tendency of the universe to change in the direction of increasing disorder. Such a definition is of course meaningless unless one defines order and disorder of the universe, and proves that disorder always increases.

Another popular definition of the Second Law which originated from Clausius himself is that the entropy of the universe always increases, and will eventually reach a maximum when the universe reaches an equilibrium state. I shall return to this topic in Section 3.12 in connection with some misuses of the entropy. Here, it is sufficient to say that the entropy of the universe is not defined either experimentally, or theoretically. Therefore, any statement regarding the entropy of the universe is meaningless.

1.8 Conclusion

In this chapter, we described three different definitions of entropy and the Second Law. All three "converge" to the same value of entropy whenever entropy can be calculated either theoretically or experimentally. My personal preference is in favor of the definition based on the SMI. The reasons for this reference are described below. Before doing this, I will briefly describe the definition of entropy based on the SMI, and the relationship between the SMI, entropy, and the Second Law.

We start by defining the SMI on the distribution of locations and velocities of a system of simple particles (i.e. having no internal degrees of freedom). We can define such an SMI for small or large systems, and for systems that are not at equilibrium.

The next step is to apply this SMI to a system of very large numbers of particles, then take the maximal value of this SMI, multiply by a constant and get the entropy of an ideal gas.

Thus, while entropy is a special case of SMI, the SMI is *not* entropy. Therefore I suggest refraining from referring to SMI as either entropy, informational entropy, or Shannon's entropy.

In both the preface and the introduction to this chapter, I claimed that the definition based on the SMI is superior to the other definitions of entropy. Here is my explanation:

1. Since entropy is a special case of the SMI, it follows that whatever interpretation you choose for the SMI (average uncertainty, average likelihood, or a measure of information), the same interpretation holds for entropy. As we shall discuss in Chapter 2, this interpretation of entropy is the only valid, solid and proven one.
2. This definition leads to an exact relationship between the SMI and the probability (Pr) of finding a specific distribution. The same relationship applies also for the entropy once we limit ourselves to equilibrium.
3. This definition removes any mystery associated with entropy. One does not need to "invent" interpretations which are based on how the system of particles look to us (order-disorder, chaos, spreading energy, etc.).
4. This definition also shows the limitations on the applicability of entropy and the Second Law. Hence, one would be careful not to apply entropy to systems for which it is inapplicable, such as living systems and the entire universe.
5. In the procedure of obtaining entropy from the SMI, we saw how the maximum SMI leads to the *uniform* distribution of locations (in absence of an external field). Also, it shows why we get the Maxwell-Boltzmann distribution of velocities at equilibrium, as well as the Boltzmann distribution of energies [For details, see Ben-Naim (2008, 2012)].
6. Finally, and most importantly, this definition shows clearly why entropy is *not* a function of time, and why it is incorrect to say that entropy tends to increase. It is the SMI of the system that can be a

function of time, and can increase with time and reach a maximum value at equilibrium.

Entropy is the value of the SMI at equilibrium. As such, it is not a function of time, it does not increase with time, and it does not reach a maximum value at equilibrium. Entropy *is* the maximal value of the SMI of a thermodynamic system.

Failing to distinguish between the SMI and entropy inevitably leads to awkward statements such as "entropy is the maximal value of the entropy at equilibrium," or that "entropy tends to increase with time and reaches a maximum at equilibrium."

Having done with the advantages of the SMI-based definition of entropy, I would like to conclude this chapter with a bold suggestion to reformulate the *Second Law* based on *probability*, rather than on *entropy*. I am well aware of the fact that most scientists view entropy as the core concept of the Second Law. Moreover, most scientists ascribe to entropy itself the power to drive all processes in the universe, not only processes in well-defined thermodynamic systems. Therefore, my suggestion to replace entropy by probability in formulating the Second Law might be viewed as heretical as, say, a suggestion to remove energy from the formulation of the First Law.

Here is my suggestion. I will formulate it for classical systems, and for systems having a very large number of particles.

First, we need to distinguish between two "levels" or probabilities. The first is the probability of finding a specific configuration of all the particles. By configuration, I mean the locations and momenta of all the particles in the system. For simplicity, we assume that there is a finite number of configurations (i.e. we already took into account the uncertainty principle and the indistinguishability of the particles). A specific configuration is a vector $c = x_1, x_2, \ldots, x_N, p_1, p_2, \ldots, p_N$, where x_i is the location vector and p_i is the momentum vector of particle i.

We define the probability $p(c)$ of finding a specific configuration c. (In a continuous case, we shall refer to $p(c)$ as the probability density.)

For each macroscopic system we define the *probability* of finding the probability distribution $p(c)$. We denote this probability by $\Pr[p(c)]$. Thus, Pr is a function of the probability distribution (for the continuous case, Pr is a *functional* of the probability density $p(c)$.)

Starting with a system with any arbitrary distribution $p(c)$, we can define both Pr and the SMI on this distribution. There is *one* distribution which maximizes both Pr and SMI. This distribution is referred to as the *equilibrium distribution* denoted by $p^*(c)$.

The formulation of the Second Law is as follows:

Starting from any constrained equilibrium state of well-defined macroscopic system, when removing the constraint, the system will move to a new equilibrium state having a new distribution $p^{**}(c)$. At this new equilibrium state, the probability $\Pr[p^{**}(c)]$ is overwhelmingly larger than the probability $\Pr[p^*(c)]$. Note that both probabilities $\Pr[p^*(c)]$ and $\Pr[p^{**}(c)]$ pertain to the system *after* the removal of the constraint.

So far we have not mentioned entropy. Entropy enters into the formulation only when the process is carried out in an *isolated system* (E, V, N constants). In such a system, the ratio of $\Pr[p^{**}]/\Pr[p^*]$ is related to the *difference in the entropy* of the system in the final and the initial equilibrium states. In this case the change in entropy must be positive.

At this point, we recognize the advantage of the probabilistic formulation of the Second Law which is valid for processes in any well-defined system. The entropy formulation is valid only for an isolated system.

If the system is characterized by T, V, N, then whenever we remove a constraint from a constrained equilibrium state, the ratio of the probabilities $\Pr[p^{**}]/\Pr[p^*]$ is related to the *difference in the Helmoltz energy* of the system, which must be negative.

If the system is characterized by T, P, N, then whenever we remove a constraint from a constrained equilibrium state, the ratio of the probabilities $\Pr[p^{**}]/\Pr[p^*]$ is related to the *difference in the Gibbs energy* of the system, which must be negative.

Thus, we see that the probabilistic formulation of the Second Law is much more general and it applies to any thermodynamic system: $(E, V, N), (T, V, N)$, or (T, P, N). The *"driving force"* for moving from one equilibrium state to another is probability. As a *result* of this process, the entropy change will be positive for an (E, V, N) system, the Helmholtz energy change will be negative for a (T, V, N) system, and the Gibbs energy change will be negative for a (T, P, N) system. Thus, we see that the probability is the central primary concept in formulating the Second Law. The entropy, Helmholtz energy, and Gibbs energy hold a secondary importance, and they are relevant to the specific systems characterized by specific thermodynamic variables.

2

Interpretation and Misinterpretations of Entropy

2.1 Introduction

Ever since the concept of entropy was introduced, people have sought a simple and intuitive interpretation of its meaning. Many interpretations were suggested over the years, such as disorder, mixing, chaos, spreading, ignorance, and freedom. Unfortunately, none of these have been proven to be a correct interpretation of entropy.

A typical procedure to arrive at an interpretation is to look at, or better yet, to imagine, a spontaneous process, then try to describe what has happened on a molecular level. If a descriptor seems always to change in the same direction as entropy, then we can conclude that this descriptor is also an interpretation of entropy.

This way of reasoning seems to be quite abstract. Therefore, I will try to clarify the issue by following a simple example: a spontaneous expansion of an ideal gas in an isolated system (Figure 1.2a).

Remember that thermodynamics was developed without any reference to the atomic nature of matter. However, knowing what happens on a molecular level, we can describe this process as follows:

1. The system evolved from a more ordered to less ordered state.
2. The energy of the system has spread from a smaller volume to a larger volume.
3. The information we have on the locations of the particles has decreased. Equivalently, the missing information, or the uncertainty about the locations, of the particles has increased.
4. The particles enjoy more freedom upon removing the partition.

All of these are qualitatively acceptable descriptions of what happens in this particular process. The first, the older one, was expressed by Boltzmann. The second seems to be first expressed by Guggenheim, and the third was explicitly stated by Lewis. The last one was suggested by Nordholm. Of course, the list of descriptors does not end here. One can view the final state as more natural, more harmonious, or more beautiful than the initial state.

The main question discussed in this chapter is not about the validity of these descriptions of the state of the system, but rather their usage as descriptors of entropy. It seems that most textbooks that attempt to explain entropy start from a description of what happens in a spontaneous process, say expansion, then correctly notice that one of these descriptors — disorder, spreading, missing information, etc. — *correlates* with the change in entropy. From these correlations, one concludes, incorrectly, that one of these descriptors is also a descriptor of entropy. This practice is quite widespread. It started with Clausius' introduction of the concept of entropy, and the ensuing endeavor continues until today. The most extreme case of such a correlation is between the direction of time, and the direction of the change in entropy. Although time is not a descriptor of the state of the system, many scientists rushed to suggest that entropy actually defines time (see Section 3.13).

None of the qualitative descriptors mentioned above can be proved to be a descriptor of entropy. Yet, curiously enough, most textbooks still use one of these to describe, sometimes even to define, entropy. In my opinion, this practice is more mysterious than entropy itself.

In this chapter, we shall discuss some of the most common interpretations of entropy. Before doing so, it is appropriate to pause and reflect on the question "Why does entropy have so many interpretations — more than any other concept in physics?" I believe that the main reason for this is that entropy has been, and still is, considered to be a mysterious quantity — some would say the most mysterious quantity — in physics. We shall discuss this aspect of entropy in Section 2.2. In the subsequent sections, we shall discuss each interpretation separately. In each case, we will present at least one example or an argument to debunk or invalidate the discussed interpretation. Finally, we show that the only interpretation which can be *proved* to be correct, and which is valid for any system for which entropy is definable, is the one based on the SMI. I hope this interpretation will bring an end to the incessant and relentless search for an intuitive interpretation of entropy. It will also remove, hopefully forever, the mystery which has enshrouded entropy for over a hundred years.

2.2 What are the Sources of the Mystery?

In my opinion, there are several reasons that "conspired" in making entropy the most mysterious concept in physics. I will list in this section what my opinion is on the main reasons. The reader is welcome to suggest others which I will be glad to incorporate in a future edition of this book.

2.2.1 *The Very New Word "Entropy"*

Everyone is familiar with concepts like force, energy, work, and the like. When you learn physics, you encounter the same words, although sometimes they have quite different meanings than the ones you are used to in everyday life. The amount of "work" that I have expended in writing this book is not measured in the same units of work (or energy) that are used in physics. Likewise, the "pressure" exerted on a politician to push for a specific law or bill is not the same as the pressure used in physics. Nevertheless, the precise concepts of "work" and "pressure" and many others, as defined in physics, retain some of the qualitative flavor of the meaning of these words as used in daily life. Therefore, it is not difficult to accommodate the new and more precise meaning conferred on familiar concepts such as force, pressure, or work. When you encounter, for the first time, a new word such as "entropy," it conjures up an air of mystery; it has a strange and uneasy effect on you. If you are not a student of physics or chemistry, and by chance hear scientists talking about entropy, you will certainly feel that this concept is beyond you and *a fortiori* so, when you hear the scientists themselves referring to entropy as a mystery.

So why did Clausius choose this particular word? Here is how Clausius explains his choice:

> I prefer going to the ancient languages for the names of important scientific quantities, so that they can mean the same thing in all living tongues. I propose, accordingly, to call S the **entropy** of a body, after the Greek word "**transformation**." I have designedly coined the word entropy to be similar to **energy**, for these two quantities are so analogous in their physical significance, that an analogy of denominations seems to be helpful.

I believe that Clausius erred in his choice of the word *entropy*, first because, entropy is not "transformation," and second, because entropy and energy are *not* analogous in their physical significance. If you open any dictionary you will find the following definitions of the word *entropy*.

In the Merriam-Webster Collegiate Dictionary (2003), entropy is defined as "change," "literary turn," "a measure of the unavailable energy in a closed thermodynamic system," "a measure of the system's degree of order."

In Yahoo's online dictionary, one finds the following:

1. The amount of thermal energy not available to work
2. A measure of the loss of information
3. A measure of disorder or randomness

In Merriam-Webster online:

1. A measure of the unavailable energy in a closed thermodynamic system
2. A measure of the system's disorder
3. The degradation of matter and energy in the universe to an ultimate state of inert uniformity

In modern Greek, entropy means "turn into," or "turn to be," or "evolves into."

$$\varepsilon\nu = in\ \tau\rho o\pi\eta = trope = transformation$$
$$\varepsilon\nu\tau\rho o\pi\iota\alpha = entropy = \text{transformation inwards}$$

All these are perhaps correct meanings of the *word* "entropy" but none is the correct meaning of the *concept* of entropy.

Cooper (1968), quoting the above paragraph regarding Clausius' explanation of his choice of word, writes:

> By doing this, rather than extracting a name from the body of the current language (say: **lost heat**), he succeeded in coining a word that means the same thing to everybody: **nothing**.

I generally agree with Cooper, but I have two additional comments.

First, I agree that "entropy" means the same thing to everybody, *nothing*. But more than that, entropy is also a misleading term. Second, I do not agree with Cooper's apparently casual suggestion that "lost heat" might be a better choice, as much as the more common "unavailable energy" attached to entropy in most dictionaries.

In fact, both the "heat loss" and "unavailable energy" may be applied under certain conditions to $T\Delta S$ but not to entropy. The reason it is applied to S rather than to $T\Delta S$, is that S, as presently defined, contains the units of energy and temperature. This is unfortunate. If entropy would have been recognized from the outset as a measure of information, or uncertainty, then it will be dimensionless, and the burden of carrying the units of energy will be transferred to the temperature T.[1] In conclusion, I believe that the very choice of the inappropriate term *entropy*, is part of the reason why entropy has become so mysterious and misunderstood.

2.2.2 *Entropy Always Increases*

In chemistry and physics, we encounter quantities which are conserved. Matter is not created from nothing, and energy is not "given us, free." We learn that both matter and energy are conserved.[2] Suddenly, we hear that there is a new quantity which is defined in terms of energy divided by temperature, which always increases. Why? If entropy is something substantial, then it seems that it will defy the conservation principle. On the other hand, if it is an abstract quantity, then how could we measure it by transferring heat (energy) into a system? I remember how I was puzzled for a long time after learning thermodynamics by this property of entropy. What is that mysterious "agent," that propels entropy upwards? Then, I heard in one of the lectures that when argon dissolves in water, the entropy of water *decreases* (see also Section 3.7). The lecturer explained that such a decrease is possible; the ever increasing entropy is only true for a spontaneous process in an isolated system.

This explanation has deepened my bewilderment. If nothing is allowed to interact with the (isolated) system, where does the system derive its energy or power from, or what causes the increase in entropy? Thus, it seems to me that the very statement that entropy always increases, without explaining why, is sufficient in mystifying the student who encounters this term for the first time.[3]

2.2.3 Saying that Entropy is a Mystery, Renders Entropy a Mystery

The very fact that many authors writing on entropy *say* that entropy *is* a mystery, *makes* entropy a mystery. This is true for writers of popular science, as well as writers of serious textbooks on thermodynamics.

Take for example a recent book, brilliantly written for the layman by Brian Greene (2004). He writes:

> And among the features of common experience that have resisted complete explanation is one that taps into the deepest unresolved mysteries in modern physics, the mystery that the great British physicist, Sir Arthur Eddington, called the arrow of time.

It seems to me that the above quoted sentence contributes to perpetuating the mystery that is no longer there. In a few more sentences, Greene could have easily explained entropy, as he explained so many other concepts of modern physics. Yet to me, it is odd that he writes: "…the deepest unresolved mysteries in modern physics," when I believe he should instead have written "Today, the mystery associated with the Second Law no longer exists."

2.2.4 Entropy, as the Almighty "Driving Force" Underlying All Processes

In my opinion, the fact that many writers on entropy and the Second Law ascribe to entropy "supernatural" powers, is the most potent contributor to the mystery associated with entropy. It is not uncommon to find the expression "ravages of entropy," ascribing to entropy the power of ravaging everything, from the splattering of an egg, to the decay and death of living systems, to the eventual "thermal death" of the entire universe. Such exaggerated and overrated "powers" ascribed to entropy undoubtedly leaves the readers of popular science books with an awesome feeling. Not only physical powers, but also our thoughts, feelings, and creativity are also attributed to the Second Law.

Here is a classic example. Atkins' book (1984) *The Second Law* starts with the following words:

> No other part of science has contributed as much to the liberation of the human spirit as the Second Law of Thermodynamics. Yet, at the same time, few other parts of science are held to be recondite. Mention of the Second Law raises visions of lumbering steam engines, intricate mathematics, and infinitely incomprehensible entropy.

I definitely do not agree with all the three quoted sentences. The first sentence is ambiguous — I fail to understand what the Second Law has to do with "liberating the human spirit." In my opinion, this is nothing but a plain, meaningless statement. The next two sentences are explicitly discouraging — "an infinitely incomprehensible entropy" does not whet your appetite to even try to taste this dish.

Here are more over-exaggerated statements made by Atkins (2007):

> The second law is one of the all-time great laws of science, for it illuminates why anything, from the cooling of hot matter to the formulation of a thought, happen at all.
>
> The second law is of central importance ... because it provides a foundation for understanding why any change occurs ... the acts of literary, artistic, and musical creativity that enhance our culture.

How could a lay reader not be deeply impressed with such statements attributed to the mighty Second Law? Such statements and many others certainly create the impression that entropy and the Second Law control everything and anything under the sun, and in the entire universe. It is even presented as more powerful than God the almighty. After all, we believe that God gives us freedom of choice, of thoughts and feelings, and of course, the choice of believing, or not believing in Him. All these freedoms have been usurped by the even-more-powerful-than-God, mysterious entropy.

Here is yet another over-the-top statement regarding the Second Law. Kafri and Kafri (2013) describe in their book:

> Why do we want more and more money regardless of how much we already have? Why do we hate to be manipulated and to lose? Why do twenty percent of the people own eighty percent of the wealth? Why in most languages, the most common word appears twice as often as the second most common word? Why the digit "1" appears in company balance sheets six and a half times more often than the digit "9?" Why does nature hate bubbles?
>
> The cause for all these phenomena is the very same law that makes water flow from high to low, and heat — from hot place to a cold one. This law, which for historical reasons is called the Second Law of Thermodynamics, states that there is a never-decreasing and always increasing quantity called "entropy."

I believe that such over-exaggerated and unfounded statements only helps to propagate the mystery associated with entropy.[4]

2.2.5 *The Multiple Interpretations of Entropy*

Any reader of popular science books must be confused when different authors offer different interpretations of entropy and the Second Law. Sometimes an author might offer several, different interpretations which are not necessarily consistent with each other. In a recent book by Lemons (2013), we find in the glossary the following:

> Entropy: A measure of the irreversibility of the thermodynamic evolution of an isolated system. Also, a measure of the spread in phase space, the uncertainty, or the missing information of the macrostate

of the system. An additive measure of the number of configurations available to a system, or, equivalently, an additive measure of the number of possibilities open to a system.

Clearly, "irreversibility" has nothing to do with entropy (see also Section 2.10). Also, it has nothing to do with "spread in phase space," or with "number of possibilities." It seems to me that the author has tried to "play it safe," and has given credit to other authors who propagate different interpretations of entropy. I must add to the author's credit that he eliminated the "disorder" interpretation from the list quoted above.

Besides the mystery propagated by many authors, there is truly a misunderstanding of entropy. As I have discussed in Chapter 1, one can use the term "entropy" as defined by Clausius without ever being bothered by the question of whether entropy has, or has not, a simple, intuitive meaning. Add to this, the famous quotation from von Neumann, who suggested to Shannon to name his "measure of information" entropy, saying:

> You should call it entropy, for two reasons. In the first place your uncertainty function has been used in statistical mechanics under that name. In the second place, and more important, no one knows what entropy really is, so in a debate you will always have the advantage.

On von Neumann' suggestion, Denbigh (1981) comments, "In my view von Neumann did science a disservice!" adding:

> There are, of course, good mathematical reasons why information theory and statistical mechanics both require functions having the same formal structure. They have a common origin in probability theory, and they also need to satisfy certain common requirements such as additivity. Yet, this formal similarity does not imply that the functions necessarily signify or represent the same concepts. The term "entropy" had already been given a well-established physical meaning in thermodynamics, and it remains to be seen under

what conditions, if any, thermodynamic entropy and information are mutually inconvertible.

I agree with the first sentence by Denbigh, and indeed, in my view, von Neumann *did* do science a disservice.

My reason for embracing Denbigh's statement is different from his. The reason is that the SMI is not entropy. Entropy is a special case of the SMI. Furthermore, Denbigh confuses "information" with the SMI.

The term "information" is far more general than the measure defined by Shannon. This is probably the reason Shannon sought the advice of von Neumann for an appropriate name for the quantity he defined to measure information (or choice, or uncertainty). It would have been helpful if an appropriate name could have been found that had retained the qualitative properties of information on one hand, yet was restricted for that specific quantity defined by Shannon, on the other hand.

2.2.6 The "Reversal Paradox" and the "Recurrence Paradox"

Perhaps the origin of the deepest mystery associated with entropy and the Second Law is to be found in the so-called Boltzmann H-Theorem, and the ensuing criticism which are known as the "reversal paradox" and the "recurrence paradox."

These apparent paradoxes create a very profound mystery associated with the Second Law. On one hand, the motions of the particles are governed by laws which are symmetrical with respect to time. On the other hand, it seems that the Second Law singles out a preferred direction of time.

Here is what Greene (2004) writes on this matter:

> We take for granted that there is a direction in the way things unfold in time. Eggs break, but do not unbreak; candles melt, but they don't

unmelt; memories are of the past, never of the future; people age, they don't unage.

Then he adds:

The accepted laws of Physics show no such asymmetry, each direction in time, forward and backward, is treated by laws without distinction, and that's the origin of a huge puzzle.

How can it be that a system consisting of particles, each obeying the time-symmetrical laws individually, but as a whole behaving differently, i.e. defying the time symmetry laws?

To answer this question, I cannot do any better than George Gamow in his book *Mr. Tompkins in Wonderland* (1940):

When he saw his glass of whisky, suddenly and spontaneously, boiling in its upper part, with ice cubes forming on the lower part, the professor knew that this process, though extremely rare, can actually occur. He might have been puzzled to observe such a rare event, but he did not look for someone playing backwards the "movie" he was acting in.

The liquid in the glass was covered with violently bursting bubbles, and a thin cloud of steam was rising slowly toward the ceiling. It was particularly odd, however, that the drink was boiling only in a comparatively small area around the ice cube.

'Think of it!' went on the professor in an awed, trembling voice. 'Here, I was telling you about statistical fluctuations in the law of entropy when we actually see one! By some incredible chance, possibly for the first time since the earth began, the faster molecules have all grouped themselves accidentally on one part of the surface of the water and the water has begun to boil by itself.'

'In the billions of years to come, we will still, probably, be the only people who ever had the chance to observe this extraordinary phenomenon.' He watched the drink, which was now slowly cooling down. 'What a stroke of luck!' he breathed happily.

In connection with the "reversibility paradox," discussed in Section 1.3.3, we can add here that many authors still propagate the conflict between the reversibility of the equations of motion and the irreversibility in thermodynamics. From the point of view of the lay reader, this leads to an unavoidable mystery.[5]

2.2.7 The Unwarranted Application of Entropy to Life and the Entire Universe

Some authors claim that life phenomena defy the Second Law (life is viewed as ordering, structuring, and organization, while entropy is viewed as a measure of disorder). On the other hand, other authors claim the exact opposite: life is created by entropy, and that entropy can explain life phenomena.

All these are obviously pure, high-grade nonsense. I have discussed this topic in great detail in Ben-Naim (2015a). I will discuss further the uses of entropy and the Second Law in connection with life and the entire universe in Chapter 3. I bring up this topic here only as one of the reasons which contribute to the mystery of entropy: Can entropy both create life and destroy it? Can life be a constant struggle against the Second Law, and at the same time be created by the Second Law?

Nonsense, is the answer!

2.3 A Necessary Criterion a Descriptor of Entropy Must Satisfy

In the following sections of this chapter, we shall discuss several interpretations and descriptors of entropy. In each of these, we will show one or more examples which refute the claim that a specific descriptor is a valid descriptor of entropy. In this section, we present a stringent criterion that any description of entropy must obey in order to be a valid descriptor of entropy.

Consider the process shown in Figure 2.1. In this process we increase the volume from V to $2V$, keeping E and N constants. We can easily claim that the disorder of the system increases. Initially, all the particles were spread over a smaller volume V. Now, they are

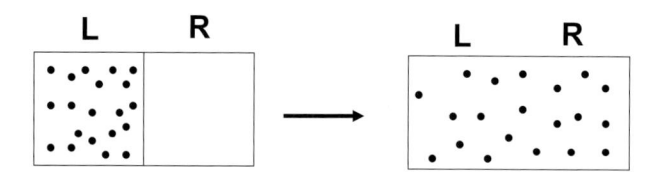

Fig. 2.1. Expansion of an ideal gas.

spread over a larger volume V. Qualitatively, we can view the final state as being more disordered than the first. Also, we can say that the larger the volume into which the gas expands, the larger the change in disorder of the system. This dependence of the disorder on the volume does not guarantee that this descriptor will satisfy the Second Law.

To satisfy the Second Law, we must also show that this descriptor must have a *negative curvature* — or is concave downwards — i.e. the slope of the entropy as a function of the volume must decrease with V.

Because of the central importance of this property of the entropy function (i.e. downward concavity), we shall devote the rest of this section to a detailed elaboration of it by means of three simple experiments.

Consider the following three simple spontaneous processed depicted in Figure 2.2. In all three processes we have two compartments, each initially having a volume V, energy E, and N particles. For simplicity we assume that the particles do not interact with each other so that we have an ideal gas system. Initially, the partition separating the two compartments is immovable, impenetrable to particles, and thermally insulating. We now do three experiments, as described in the following.

2.3.1 *Volume Exchange Between Two Compartments*

In this experiment, we replace the partition by a movable partition (see Figure 2.2a). Since the two subsystems are identical, the replacement

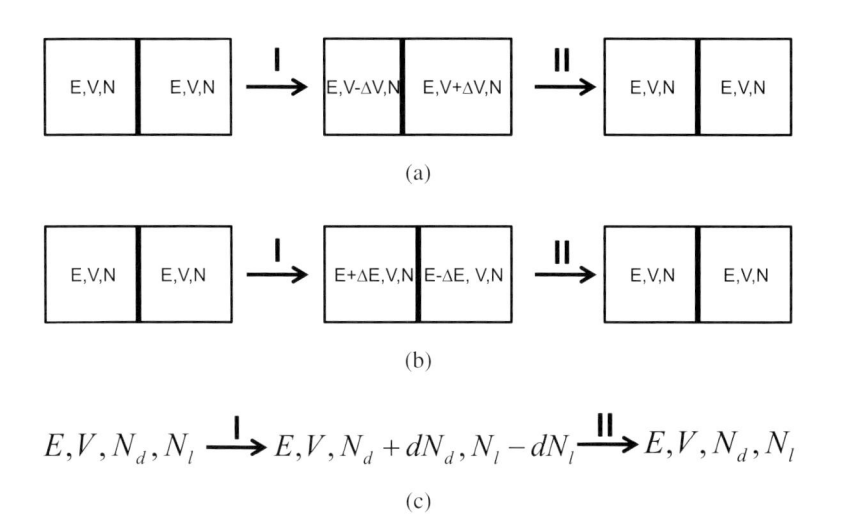

Interpretation and Misinterpretations of Entropy | 117

(a)

(b)

$$E, V, N_d, N_l \xrightarrow{\text{I}} E, V, N_d + dN_d, N_l - dN_l \xrightarrow{\text{II}} E, V, N_d, N_l$$

(c)

Fig. 2.2. Three experiments in an isolated system.

of the partition by a movable one will not cause any observable change in the entire system. We know that the position of the partition might fluctuate, but these fluctuations will be extremely small and unnoticeable. Next, we move the partition from its location at x, to either the right or to the left, and then release it. This process is indicated I in Figure 2.2a. The process we want to study next is indicated by II in Figure 2.2a. We can calculate the exact change in entropy in this process (II), but we do not need the exact amount of the change in entropy; all we need to know is the *sign* of the change in entropy, which in this process (II) must be positive.

Now, consider any of your favored descriptors D. When we move the partition to say, the left, we have decreased the volume of the left compartment (ΔV) and increased the volume of the right compartment, by the same amount ΔV. Next, we release the partition, so it will move spontaneously to its original position. We have assumed that the descriptor D is a monotonic increasing function of the volume. Therefore, when we release the partitions, the value of D must increase in the left compartment and decrease in the right

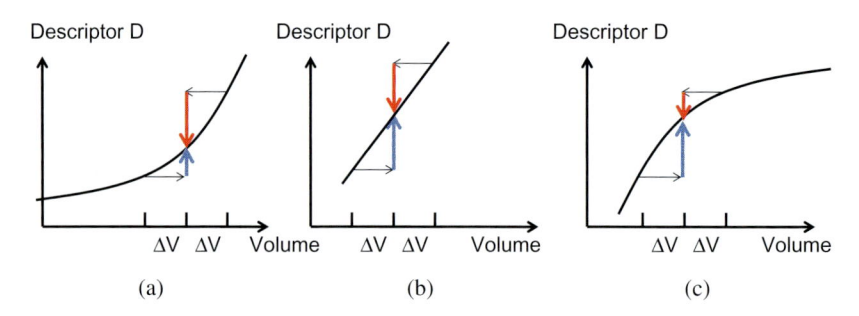

Fig. 2.3. Three possible curves of a descriptor of entropy as a function of volume.

compartment. Three possible curves are shown in Figure 2.3. If the dependence on D on the volume is *linear* (Figure 2.3b), then the *net change* in D in the spontaneous process (II) will be zero (the length of the blue arrow is equal to that of the red arrow). If the curve is concave upward (as in Figure 2.3a), i.e. the curvature is positive (the slope increases with the volume), then the net change in D is negative. The third case is when the curve is concave downwards (Figure 2.3c) — the slope *decreases* with the volume. In this case, the net change in entropy is *positive*. In each of the curves in Figure 2.3, we show the change in D for the right and the left compartments in colored arrows.

We can conclude that if the change in D is expected to be in the same direction as the change in entropy, it must be a concave downward function of the volume of the system. It is only in this case that the net change in D will be positive for the spontaneous process II described in Figure 2.2a. In other words, the blue arrow in Figure 2.3 must be larger than the red arrow which means that the net change in D is positive as required by the Second Law.

2.3.2 Energy Exchange Between Two Compartments

In this experiment we start with the same initial state of the entire system. We transfer an amount of energy ΔE from one compartment to another, say, by heating the left compartment and cooling the right compartment, process I in Figure 2.2b. The way we do this is not important; what matters is that the net result of this transfer of energy is that the energy of the left compartment has increased by the amount ΔE, and the energy of the right compartment by the amount of $-\Delta E$; the volumes and the number of particles in each compartment did not change. Note that if the compartments contain ideal gases, then the energy of the system is the sum of the kinetic energy of its particles. In this particular case the change in energy is equivalent to the change in the temperature of the system, or of the velocity distribution of the particles in each compartment.

After we have transferred the energy, we replace the partition between the two compartments by a heat-conducting (diathermal) partition. What will happen? The process we are considering is indicated by II in Figure 2.2b.

As you can guess, heat will flow from the high-temperature to the low-temperature compartment until thermal equilibrium is reached. The temperature on the two sides of the partition will be equal. You can easily calculate the change in the entropy in this spontaneous process and find that it is positive.[6]

Can you argue that your favored descriptor D has also increased in this process? If you have convinced yourself that the quantity D increases monotonically with the increase of the energy of the system, then you can conclude that the D value of the left compartment must decrease while the D value of the right compartment must increase. Can you show that the *net change* of D in the spontaneous process in the combined system has also increased? For this to occur, you must have a concave downward dependence of D on the energy E of the system, i.e. a similar curve shown in Figure 2.3c, but for D as a function of E.

2.3.3 *Material Exchange Between Two Compartments*

As an exercise, construct an experiment similar to the previous examples. The initial state is as before, but now we transfer ΔN particles from the right to the left compartment. For instance, we can add ΔN

to the left and extract ΔN from the right. What is important is that the net change is such that the number of particles on the right has decreased, and on the left it has increased by the same amount, ΔN.

Note that whenever we transfer particles we also transfer their kinetic energy. Therefore, when we remove the partition between the two compartments, both the particles and their kinetic energy will be transferred from one compartment to the other.

A challenging exercise

Can you think of a spontaneous process in which molecules "flow" from one state to another with a net change in entropy, but no change in volume and no change in energy? For the answer, see Section 3.9. A hint is given in Figure 2.2c.

To conclude, we designed three experiments where in each a spontaneous process occurs in an isolated system. In each of these processes, the entropy increases. Any suggested descriptor of entropy D must do likewise.

To the best of my knowledge, no one has ever shown, not even qualitatively, that any of the common descriptors of entropy fulfills this necessary property of entropy. The only interpretation of entropy which fulfills this criterion is the one based on the SMI.

2.4 The Association of Entropy with Disorder

Perhaps the oldest metaphor that has endured for the longest time, and is the most popular metaphorical description of entropy, is *disorder*. You can find this interpretation of entropy even in most recent popular books by most well-known authors. See also Ben-Naim (2015a, 2016a).

Boltzmann was probably the first to associate the Second Law with increasing disorder. Here are some quotations.[7]

> ... the initial state of the system ... must be distinguished by a special property (ordered or improbable) ...

> ... this system takes in the course of time states ... which one calls disordered.
>
> Since by far most of the states of the system are disordered, one calls the latter to the probable states.
>
> ... the system when left to itself, it rapidly proceeds to disordered, most probable state.

First, note that at least from these quotations, Boltzmann uses the term "disorder" to describe what happens: "when (the system) is left to itself, it rapidly proceeds to disordered, most probable state." It does not seem that Boltzmann equated entropy with disorder as many authors inferred.

At this point, I would like to pause and say something personal, which might sound paradoxical as I have criticized the "disorder" metaphor in my previous publications (2007, 2008, 2010): From what I have read from Boltzmann (through the translations of Brush (1976, 1983), I found no place where Boltzmann equated entropy with disorder. Boltzmann did describe the state of the system, or changes in the state of the system using the terms "order-disorder." If this is the case, then I do not see any room for criticizing Boltzmann for that usage of order-disorder. One could describe the final state of a system as disordered, as more spreading, more freedom, and I can add more harmonious, more perfect, more beautiful, or whatever one chooses. This is not a subject for criticism. It is a purely personal view, or a personal description of the *state* of the system. The main questions to ask are: Can *all* spontaneous processes be described as going from order to disorder? My answer to this question is: No! Some examples are provided below. Can one prove that entropy *is* a measure of disorder? My answer is definitely No! In fact, there were attempts to prove, or at least justify, the disorder metaphor. We shall discuss these below.

Second, take note that whenever Boltzmann describes what goes on in the spontaneous process, he also uses the term "probable state." In this sense, Boltzmann was absolutely right. As we have discussed in Chapter 1, the processes do occur because they move from less probable to more probable states.[8]

How did "disorder," which is a qualitative, ill-defined and highly subjective descriptor of the state of a system, become a descriptor of entropy?

Under the general term "disorder," I have included other similar descriptors such as: "disorganization," "mixed-upness," "randomness," "unstructuredness," "chaos," and many more. All these are qualitative descriptors of the state of a system. Anyone looking at a solid and a liquid would have no difficulty in distinguishing the ordered from the disordered state. But given any two ordered systems or two disordered systems, it is impossible to tell which system is more ordered. A simple example is shown in Figure 2.4. A more difficult example is when we initially have two bodies at different temperatures. After being in thermal contact, the system will reach a final equilibrium state, hence, a uniform temperature. Can you tell which of these states is more ordered than the other?

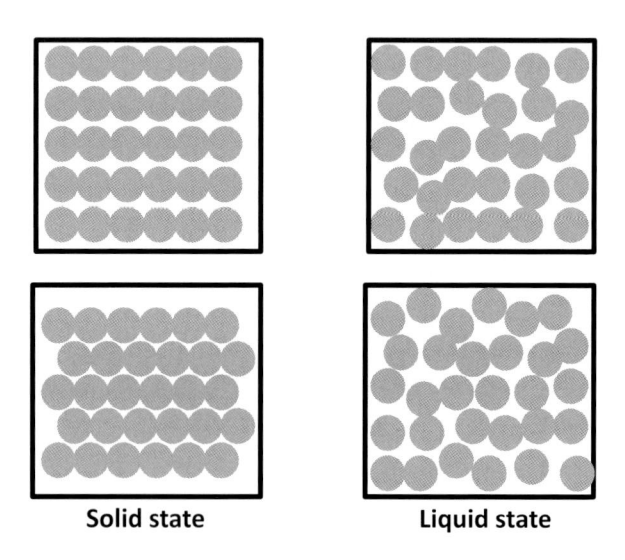

Solid state   **Liquid state**

Fig. 2.4. Comparison of two ordered and two disordered systems.

Another example is the following: suppose that we accept the "disorder" interpretation of entropy. Take two gases, the first one characterized by E, V, N and the second one by $2E, 2V, 2N$. Clearly, the entropy of the second is twice as large as the first. It is in my view difficult to claim that the second is more "disordered" than the first, i.e. that "disorder" is an extensive property. But suppose we accept this interpretation. Next I show you two perfect crystals characterized by E, V, N and by $2E, 2V, 2N$. Again, the entropy of the second system is twice as large as the entropy of the first. Can one claim that the "disorder" of the second is twice as large as the first? Perhaps, one can say that the "order" of the second is larger than the first, but not the "disorder."

Perhaps the most common reason for associating entropy with disorder is the process of mixing shown in Figure 2.5a. This is also the process of mixing studied by Gibbs which led him to identifying

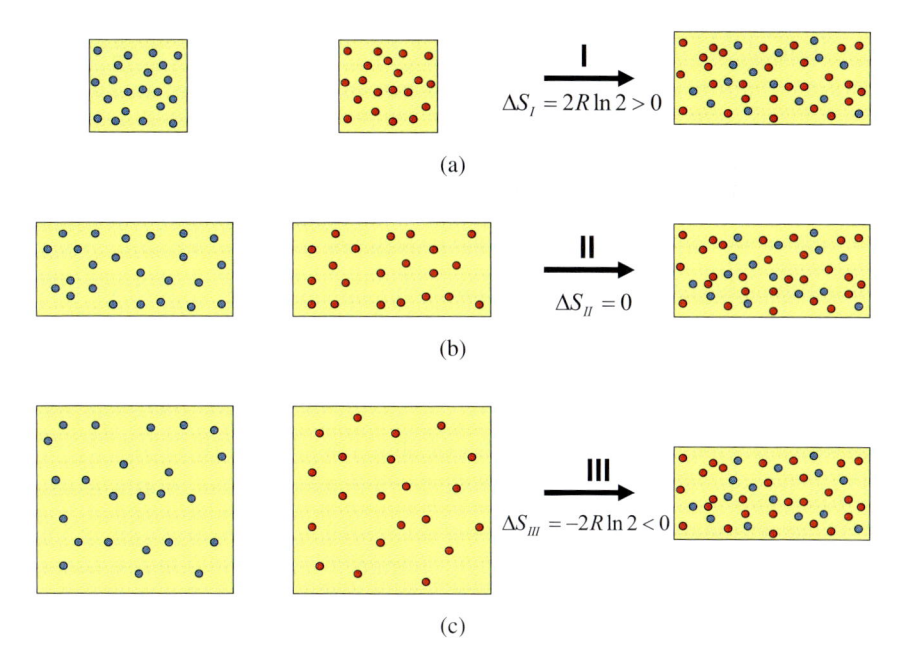

Fig. 2.5. Three processes of mixing with positive, zero, and negative change in entropy.

positive entropy changes with "mix-upness," a term quite closely related to disorder.

Indeed, if we look at the process in Figure 2.5a, we *see* mixing. We can also calculate the entropy change for this particular process of mixing two ideal gases.[9] The result is $\Delta S = 2R \ln 2 > 0$. Look at Figure 2.5b; on the left-hand side we have two boxes of equal sizes having the same number of particles. The particles in the first box are different from the particles in the second box. Now we mix the two systems into one box of the same size. Is the system on the right-hand side more disordered than the one on the left-hand side?

Everyone would agree that a mixture of two different compounds is in a less ordered state than the unmixed compounds. Since the entropy change for the mixing process in Figure 2.5a is positive, and since mixing is viewed as a process of disordering, one concludes that positive change in entropy is associated with increase of disorder.

Indeed, this conclusion is correct for this particular process of mixing (Figure 2.5a). However, it is not always true that mixing is associated with increase in entropy.

In the process shown in Figure 2.5b we have mixing. It is also intuitively clear that the system on the right-hand side is more disordered. However, the entropy of the systems on the two sides of Figure 2.5b is the same. This can be easily calculated for two ideal gases. [For details see Ben-Naim (2008, 2012).] We also note that if the particles interact with each other and if the interaction energies between the pairs AA, AB, and BB are different, then we can get both positive and negative changes in entropy in the process of the type shown in Figure 2.5b. Thus, in general one cannot conclude that mixing is a spontaneous process associated with an increase in entropy. Figure 2.5c shows a process of mixing with a negative change in entropy. This is not a spontaneous process. However, one can show that a spontaneous process can occur in which two components get unmixed, yet the entropy change is positive [see Ben-Naim (2008, 2012)].

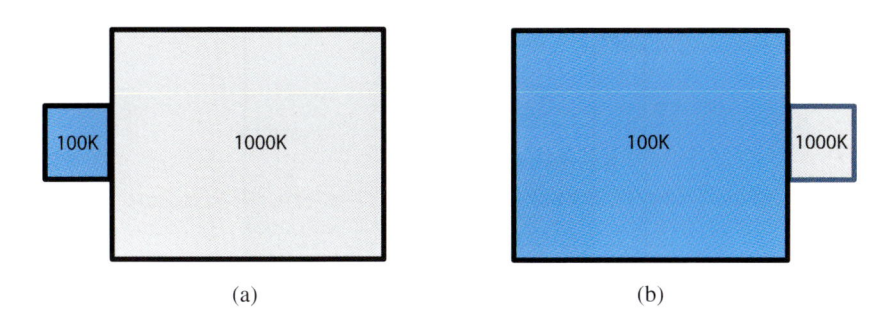

Fig. 2.6. Two experiments with ice and steam.

Pause and ponder

Figure 2.6 shows two processes of heat transfer. In Figure 2.6a, we have one mole of ice at very low temperature, say 100 K, and 100 moles of steam at very high temperature, say 1,000 K.

In Figure 2.6b we have one mole of steam at 1,000 K and 1,000 moles of ice at very low temperature, say 100 K.

We bring the two systems into thermal contact. The combined system in each case is an isolated system. Try to answer the following questions for each of the processes (a) and (b), separately. Write down your answers before you consult Note 10.

1. Describe qualitatively what will happen.
2. What will be the final temperature at equilibrium?
3. In which direction will the total entropy of the combined system change?
4. Can you tell whether the combined system will become more ordered or disordered?

One can invent many examples where it would be difficult, ambiguous, or impossible to determine which system is more ordered than the other. An example discussed qualitatively and quantitatively in Ben-Naim (2012) is the expansion of an ideal gas in a gravitational field. Figure 2.7 shows two such spontaneous expansion processes in

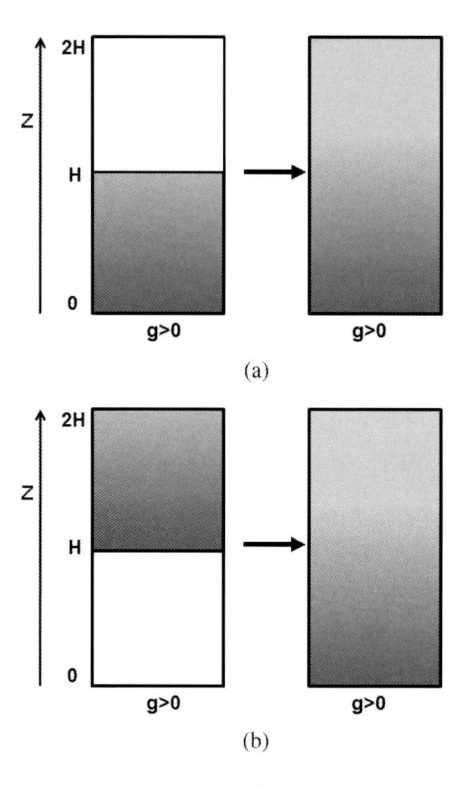

126 | Entropy: The Truth, the Whole Truth and Nothing But the Truth

Fig. 2.7. Expansion in a gravitational field (a) upward and (b) downward.

an isolated system. In both processes the entropy increases. Can you tell in which system the change in disorder is larger or smaller?

The important conclusion from the examples discussed here is that order and disorder, *sometimes*, but not always, may be used to compare two systems, or two states of the same system. However, one cannot show that in *any* spontaneous process in an isolated system, the "degree of disorder" necessarily increases. Even if we could do this there still lingers the question of relating the extent of disorder with the entropy. The answer to this question can only be given if we can define quantitatively the concept of order or disorder, and if we can show that this quantity is equal to entropy.

Yet, surprisingly most textbooks on thermodynamics still teach us that entropy is a measure of disorder. For instance, in a recent popular book, Atkins (2007) writes:

> We shall identify entropy with disorder and with disorder in mind, we shall explore the implication of Clausius' expression and verify that it is plausible in capturing the entropy as a measure of the disorder in a system.

If you read the entire book, you will not encounter any "verification" that is promised in this quotation, nor will you find a plausible definition of disorder. Other examples abound in popular science books.

In spite of the fact that "disorder" is not a well-defined concept, and in spite of the fact that "disorder" only sometimes describes qualitatively the state of a system, the concept of "disorder" as a metaphor for entropy has prevailed in the literature for over a hundred years. The only explanation I can offer for this phenomenon is that in many examples, positive changes in entropy seem to correlate with increase in disorder. I also suspect that those who claim that entropy is a measure of disorder simply do not understand what entropy is.

Not only is entropy interpreted as a measure of disorder, but also the Second Law is sometimes formulated in terms of the tendency of spontaneous processes to proceed from order to disorder. From this interpretation, it is easy to slip into the example of a child's room, or the kitchen which always becomes disordered when left untended. Another favorite example is an ordered pile of pages of a book, and the same pages of the book strewn in the room. The former is supposed to have a lower entropy than the latter (see Figure 2.8). Unfortunately, the Second Law cannot say anything about the difference in entropy when the pages of the book are scattered.

To conclude this section I should mention one "definition" of disorder appearing in Callen's book on thermodynamics (Callen, 1985).

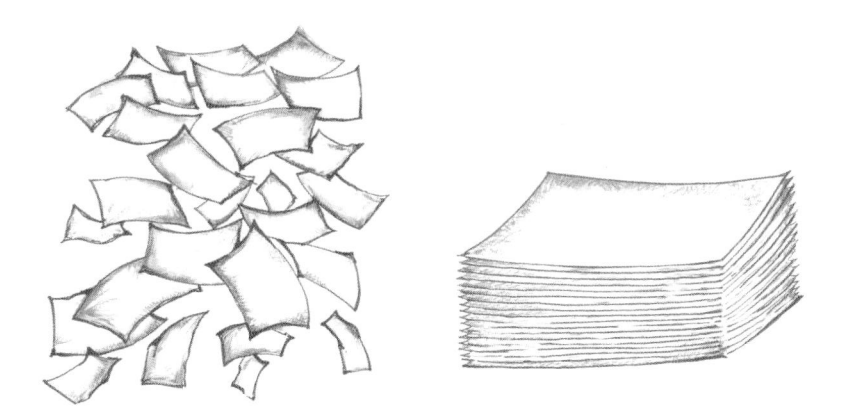

Fig. 2.8. Ordered (*right*) and disordered (*left*) pages of a book. Which side has a larger entropy?

Callen defines "Shannon's disorder" by the following:

$$\text{Disorder} = -\sum p_i \log p_i.$$

The right-hand side of this equation is a well-defined quantity. Unfortunately, Shannon never defined "disorder." What Callen has defined is not "Shannon's disorder" but "Shannon's measure of information." Therefore, the positioning of "disorder" on the left-hand side of the equation is unwarrantable and unjustifiable.

To conclude this section we can say that order and disorder can be used to describe qualitatively the state of a system. However, neither entropy nor the Second Law can be interpreted as the extent of disorder, or in terms of changes in the extent of order.

2.5 The Association of Entropy with the Spreading/Dispersion/Sharing of Energy

The second descriptor of entropy as "spreading" was probably first suggested by Guggenheim (1949). Guggenheim started with the Boltzmann definition of entropy in the form $S(E) = k \log \Omega(E)$, where k is a constant, and "$\Omega(E)$ denotes the number of accessible independent quantum states of energy E for a closed system." This is the correct definition of Ω.

Later in his article Guggenheim discusses the process of heat flow in which an increase of entropy represents "an increase in the number of accessible states, more briefly an increase of accessibility or spread."

The first part of the quotation is correct; namely, the change in entropy in the spontaneous process of heat flow represents an increase in the number of accessible states (here "states" mean "microstates"). The most important word in this sentence is *number*. In the second part of the sentence this word is deleted and what remains is "increase of accessibility or spread." Obviously, this is a deficient description of the change in Ω. Yet, it is acceptable because the two parts of the sentence are adjacent to each other, and it is clear that by "increase of accessibility" he meant "increase in the number of accessible states." However, in another part of the same article Guggenheim further shortened the description of Ω.

> To the question what in one word does entropy really mean, the author would have no hesitation in replying "Accessibility" or "Spread." When this picture of entropy is adopted, all mystery concerning the increasing property of entropy vanishes.

Here, the terms "accessibility" or "spread" are used to describe "what does entropy really mean." However, note that Guggenheim does not define either term. He simply used these two terms to describe what the "number of accessible quantum states of energy E for a closed system" is, i.e. the quantity Ω as it appears in Boltzmann's formula for entropy. As we have quoted earlier, "$\Omega(E)$ is the number of accessible independent quantum states of energy..." It is absurd to pick one word "accessibility" or "spread" to describe *entropy*. It is absurd for two reasons. First, because neither of the two words alone fully describes Ω. Second, and most importantly, Guggenheim correctly defines Ω as the "number of accessible states," but then takes a briefer description of Ω to define entropy. This is, of course untenable. Ω and $\log \Omega$ are two different quantities. They are related to each other but they do not necessarily have the same meaning. For example, in the 20Q game, Ω is the number of objects from which I choose one object, and log Ω represents the minimum number of binary questions you have to ask to find out which of the Ω object was chosen. Clearly, these are two different things, albeit related mathematically by the logarithm function. It would be deficient to use a shortened description of Ω, say, "the choice of objects" to describe Ω, but it would be utterly absurd to use "the choice of objects" to explain "the minimal number of questions to be asked in the 20-question games." Thus, it is okay to use a shortened sentence or even one word, or an acronym, as long as the full sentence is given, and that it is understood what the shortened version stands for. It is absurd to replace the full sentence by its shortened version. It is *a fortiori* absurd to teach students that entropy means "spreading," detaching the word "spreading" from the full definition of Ω.

Finally, I doubt that by adopting either "accessibility" or "spread," all the mystery surrounding entropy will vanish. In fact, such a descriptor of entropy only increases the mystery associated with entropy.

In several recent publications Lambert (1999, 2002, 2007) has advocated the interpretation of entropy in terms of spreading. His interpretation is summarized as follows:

> Entropy change is the measure of how more widely a specific quantity of molecular energy is dispersed in a process, whether isothermal gas expansion, gas or liquid mixing, reversible heating and phase change or chemical reactions.
>
> Concisely, the second law is "Energy of all types changes from being localized to becoming spread out, dispersed in space if that energy is not constrained from doing so."

Thus, in these two sentences Lambert defines both entropy and the Second Law, based on an undefined and fuzzy concept of the "spreading" descriptor. It is even more unfortunate that about 24 textbooks have changed from one ill-defined descriptor "disorder" to yet another ill-defined descriptor "spread" or "dispersal." Clearly, saying that "entropy is... dispersal of energy," does not make entropy a measure of energy dispersal. One can replace the word *disperse* by disorder, information, beauty or ugliness and the resulting statement will have the same degree of truthfulness as the two quotations above.

To justify this interpretation of entropy — or perhaps to "show" that it is equivalent to Boltzmann entropy, Lambert writes the two equalities $\frac{dq_{rev}}{T} = dS = k_B \ln \frac{W_2}{W_1}$.

Here, there are two different "definitions" of entropy change. On the right-hand side, the change in entropy is due to change in the number of states in an isolated system. The equality on the left-hand side is Clausius' definition. It pertains to a process of transferring a small quantity of heat to a system at constant temperature. These two equalities pertain to two different processes, and therefore cannot be equated. In addition, these equalities have nothing to do with any interpretation of entropy.

There are many processes, e.g. expansion of ideal gas, which could be viewed as spreading of energy. Here is an example where the "spreading of energy" cannot explain the positive entropy change. Consider the mixing of two kinds of molecules in an isolated system, Figure 2.9a. If we start with one mole of A, and one mole of B, the change in entropy in this process will be $2R \ln 2$.

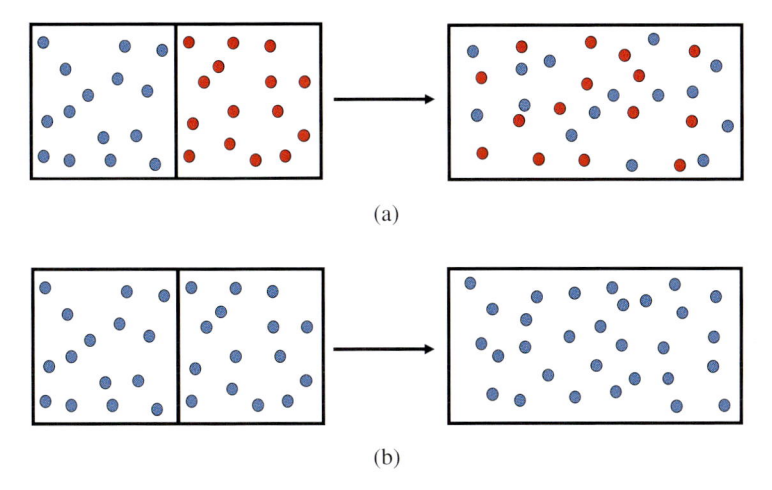

Fig. 2.9. (a) Mixing of two different kinds of molecules, and (b) "mixing" the same kind of molecules.

For simplicity, we assume that A and B are atoms having only kinetic energies. If initially the two gases are at the same volume V and have the same energy E, then after the mixing of the two gases the energy of the molecules in each compartment will not change. It is difficult to argue that the entropy change is due to the spreading of the energy in this process.

I once wrote to someone who is in favor of the "spreading" interpretation, and asked him how he can explain the entropy change in this mixing process. His answer: The energy of *each* gas separately spreads from V to $2V$.

I then asked, if this were true, what about the process shown in Figure 2.9b where the same gas, say A, is in the two compartments. When we remove the partition, we can also say that the energy of molecules from the left compartment has spread from V to $2V$. At the same time the energy of the molecules from the right compartment has also spread from V to $2V$. How do you explain the fact that the entropy change in this process is zero? On this question, I did not get an answer. I should also add that I do not mention the person's name because he explicitly asked me not to reveal his name.

As in the case of "disorder," one can use the word "spreading" to describe either the state of the system or the change in the state of the system. For instance, in the expansion of an ideal gas one can qualitatively say that the particles, or the energy carried by the particles, have spread from a smaller volume to a larger volume. However, one cannot show that the spreading of energy in the final equilibrated system is larger than the spreading before the equilibrium. To illustrate this, one must show that the spreading function is not only monotonically increasing of the energy but is also a concave downward function of the energy.

It is interesting to note that some authors who advocate the "spreading" interpretation of entropy would also say that when the energy spread is larger, the entropy change should be larger too. Such statements lead to an amusing conclusion.

Consider the change in the entropy in the expansion of one mole of an ideal gas, from volume V to $2V$. In either an isolated system, or in an isothermal process, the change in entropy is $\Delta S = R \ln \frac{2V}{V} = R \ln 2$, where R is the gas constant. You can repeat the same process at different temperatures; as long as you have an ideal gas, the change in entropy in this process is equal to $R \ln 2$, independently of the temperature (as well as of the energy) of the gas. Those who advocate the "spreading" interpretation would also say that the larger the energy spread, the larger the "spreading," and the larger the entropy change in the expansion process. From this, it follows that the entropy change in the expansion of an ideal gas must be larger the higher temperature of the gas.

This conclusion is, of course, not true. As long as we have an ideal gas the change of entropy in the expansion of one mole of gas from V to $2V$ is independent of temperature. It is always $R \ln 2$. Yet, some authors insist that $R \ln 2$ depends on the temperature. To see why, the reader is referred to Ben-Naim (2012).

The SMI interpretation is simple and straightforward. In the expansion process we lost one bit of information per particle, hence the change in the SMI is $N \log_2 2$, which is the same as ΔS except for the Boltzmann constant k_B, and the change in the base of the logarithm. Clearly, this change in SMI is independent of temperature.

Pause and ponder

Consider the case of a racemization reaction. There are molecules having an asymmetric center (or a chiralic center) such as alanine (see Figure 2.10). These molecules have two enantiomers, each being a mirror image of the other. They are called *levo* (*l*) and *dextro* (*d*). These terms have something to do with the effect of molecules on polarized light. For our purpose, we view these two molecules as two isomers of the same chemical compound. The two isomers have the same molecular properties, yet they are distinguishable, and in fact, can even be separated.

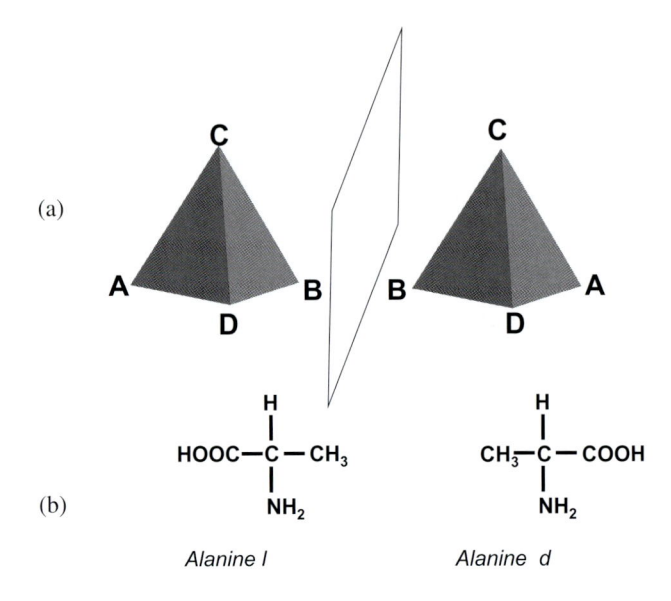

Fig. 2.10. A molecule and its mirror image — two isomers with the same molecular properties. Two enantiomers of alanine shown in (a) three dimensions, and (b) two dimensions.

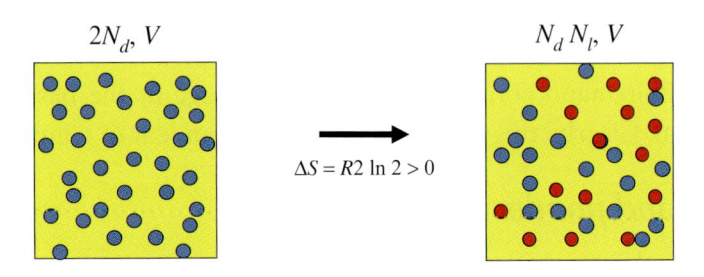

Fig. 2.11. Racemization process of pure d into two enantiomers, d and l.

Suppose we start with one mole of the pure d form in an ideal gas phase. We add a catalyst and the system evolves into a mixture of d and l forms (see Figure 2.11). At the new equilibrium state we have half a mole of the d form, and half a mole of the l form. This process is called racemization. More on this reaction in Section 3.9.

Answer the following questions:

1. Why do we always find equal amounts of *d* and *l* forms at equilibrium?
2. What do you expect the entropy change to be in this process?
3. Can you interpret, even qualitatively, the entropy change in this process, using your favorite descriptor of entropy?
4. Can you interpret the entropy change in this process in terms of spreading of energy?

The reader is urged to try answering these questions, even qualitatively, before reading Section 3.9, where a more complete discussion of this reaction is presented.

2.6 The Association of Entropy with Information

The earliest association between information and entropy is probably due to Lewis (1930):

> In the simplest case, if we have one molecule which must be in one of two flasks, the entropy becomes less by $k \ln 2$, if we know which is the flask in which the molecule is trapped.
>
> Gain in entropy always means loss of information and nothing more.

Note that this quotation dates back almost 20 years before Shannon published his measure of information. From the second quotation, it seems that Lewis is using the term "information" in its colloquial meaning: when a gas expands we have a sense of losing some locational information. However, from the first part of this quotation, it seems that Lewis is saying exactly what the SMI interpretation states, namely we lose log 2 bits per particle.

The *informational* interpretation of information is perhaps the one that had raised the most vigorous objection. The objection was fully justified as long as the concept of *information*, or a *measure of information*, was vague, not well defined, and highly subjective.

In 1948, Shannon published *A Mathematical Theory of Communication*. In this article, Shannon defined a quantity which measures the extent of information, or of missing information, or uncertainty associated with any probability distribution. In Chapter 1 we referred to this quantity as the Shannon measure of information (SMI).

Jaynes (1957) and later Katz (1967) have used the so-called principle of maximum entropy to derive the fundamental distributions of statistical mechanics. In fact, both Jaynes and Katz used the principle of maximum SMI to derive the fundamental distributions of statistical mechanics. As we have seen in Chapter 1, the thermodynamic entropy is the maximal value of the SMI when applied to certain distribution functions relevant to a thermodynamic system.

In his first paper on this subject Jaynes (1957a) wrote:

> Henceforth, we will consider the terms "entropy" and "uncertainty" as synonyms. The thermodynamic entropy is identical with information theory — entropy of the probability distribution except for the presence of Boltzmann's constant.
>
> ... we accept the von Neumann-Shannon expression for entropy, very literally as a measure of the amount of uncertainty represented by the probability distribution; thus entropy becomes the primitive concept ... more fundamental than energy.

I generally agree with the spirit of Jaynes' writing. What Jaynes referred to as "information theory entropy" is nothing but the SMI. Thus, although the two concepts of entropy and SMI have identical formal structure when defined in terms of probability distributions, they are very different in their range of applicability. Again, it should be stressed that the SMI is defined for any probability distribution, whereas entropy is defined on a limited set of distributions pertaining to a thermodynamic system at equilibrium.

Furthermore, confusing Shannon's *measure* of information and the general concept of information, has raised vigorous debates. For instance, in 1994 Gell-Mann wrote:

> Entropy and information are very closely related. In fact, entropy can be regarded as a measure of ignorance. When it is known only that a system is in a given macrostate, the entropy of the macrostate measures the degree of ignorance the microstate is in by counting the number of bits of additional information needed to specify it, with all the microstates treated as equally probable.

I fully agree with the content of this quotation by Gell-Mann, yet Prigogine (1997), commenting on this very paragraph, writes:

> We believe that these arguments are untenable. They imply that it is our own ignorance, our coarse graining, that leads to the second law.

The reason for these two diametrically contradictory views by two great scientists has its sources in the confusion of the *general* concept of information with the specific measure of information as defined by Shannon. I have discussed this issue in great detail in Ben-Naim (2015a).

In my opinion, Gell-Mann is not only right in his statement, but he is also careful to say "entropy *can* be regarded as a measure of ignorance... Entropy... measures the degree of ignorance." He does not say "our own ignorance," as misinterpreted by Prigogine. As we have discussed in Chapter 1, a given 20Q game defines a value for the SMI. The value of the SMI does not depend on who knows, or does not know, the amount of information associated with the game.

Indeed, the term "information," in its colloquial sense, might be highly subjective. However, within information theory, the SMI is not a subjective quantity. Gell-Mann uses the term "measure of ignorance" as a synonym to "measure of lack of information." As such, these are also objective quantities that belong to the system, and are not the same as "our own ignorance," which might or might not be an objective quantity. The entropy of a system has nothing to do with "our information," or "our ignorance" of the microscopic state of the system. It is not uncommon to find in the literature statements like the following: "If there are W states, the entropy is $k_B \ln W$. If we know in which specific state the system is, then the entropy is zero." This is, of course not true. The entropy does not depend on our *knowledge* of the state of the system, but on the *number* of *states* of the system!

The misinterpretation of the information-theoretical entropy as subjective information is quite common. Here is a paragraph from Atkins' preface from his book *The Second Law* (Atkins, 1984):

> I have deliberately omitted reference to the relation between information theory and entropy. There is the danger, it seems to me, of giving the impression that entropy requires the existence of some cognizant entity capable of possessing "information" or of being to some degree "ignorant." It is then only a small step to the presumption that entropy is all in the mind, and hence is an aspect of the observer. I have no time for this kind of muddleheadedness and intend to keep such metaphysical accretions at bay. For this reason I omit any discussion of the analogies between information theory and thermodynamics.

Atkins' comment and his rejection of the informational interpretation of entropy on the grounds that this "relation" might lead to the "presumption that entropy is all in the mind" are ironic. Instead, he uses the terms "disorder" and "disorganized," etc., which in my view are concepts that are far more "in the mind."

The reason for the confusion is that the term "information" itself has numerous interpretations. In the most general sense, information is any knowledge that we can obtain by our senses. It is an abstract concept which may or may not be subjective. The information on "the weather conditions in the state of New York" might have different significance, meaning, or value to different people. This information is *not* the subject of interest of information theory. When Shannon sought a quantity to measure information transmitted across communication lines, he was not interested in the *content,* or the *value,* or the *meaning* of information, but in a quantity that *measures* the *amount* of information that is being transmitted. Thus, the most important point to understand before using information theory is to recognize that "information" and the "measure of information" are two different things, just as a *circle* and its *diameter* are two different things. Furthermore, one should also recognize that there are only certain types of information that can be measured, and there are different measures that can be assigned to information. Thus, a circle, a square, or a hexagon may be assigned a measure, but an abstract concept like beauty, peace, or love cannot. For the same object, say a circle, one can assign different measures such as its diameter, its area, or the length of its circumference. Therefore, one should be very careful when one talks about "information" and "a measure of information" without specifying which information is measured, and what the measure we assign to the information is.

The fact that information can be subjective has led many to reject the informational interpretation of entropy because entropy is an objective quantity. Clearly, this rejection is a result of either misunderstanding the SMI, or confusing the SMI with the general term "information."

It is ironic that some authors who accept the informational interpretation of entropy, but confuse the SMI with information, reach the unwarranted conclusion that entropy is also a subjective quantity.

Here is a typical quotation [von Baeyer (2005)]:

> It is not just the number of ways a system can be rearranged, but more specifically the number of rearrangements consistent with the known properties of the system. Known by whom? Measured by what observer?
>
> By this reckoning entropy is not an absolute property of a system, but relational. It has a subjective component — it depends on the information you happen to have available.

Boltzmann understood this connection and made it more specific. He pointed out that since the value of entropy rises from zero we know all about a system, to its maximum value when we know least, it measures our ignorance about the details of the motions of the molecules of a system. Entropy is not about speeds or positions of particles, the way temperature and pressure and volume are, but about our lack of information.

In my view, entropy does not have a "subjective component!" Furthermore, Boltzmann did not "understand" this (erroneous) conclusion. The entropy of a system does *not* rise from zero... to maximum value. The entropy of a system is fixed for a system at equilibrium, *independently* of what we know, or do not know, about the "details of the motions of the molecules." Such statements reveal a profound misunderstanding of entropy by the author who wrote that quoted paragraph.

2.7 Does Entropy Depend on Our Knowledge About the System?

In Section 2.6, we discussed the association of entropy with information. We saw that if by information, one means the SMI associated with the distributions of a thermodynamic system, then such a usage of information (or lack of information) is justified. It is not justified when one uses the term "information" in the general, colloquial sense. The information on the color of the die has nothing to do with the SMI associated with the probability distribution of the outcomes of the die. It certainly does not have anything to do with the entropy of the die, viewed as a macroscopic thermodynamic system.

In this section, we discuss the question of whether the entropy depends on the knowledge we have (or do not have) on a thermodynamic system. We shall discuss three questions: (1) Does the entropy of the system depend on how precisely we want to specify the molecular, or microstate, of the system? (2) Does the entropy depend on how much we know on the extent of intermolecular interactions between the particles? And finally, (3) Does the entropy depend on our knowledge of the composition of the system?

2.7.1 Does Entropy Depend on the Precision We Choose to Describe the Configuration of the System?

Consider for simplicity an ideal gas in a volume V. Clearly, the number of configurations for the locations of the particles is infinite, and the corresponding SMI will be infinity. Now suppose we divide the volume V into n cells, each of volume V/n. We choose to count the number of configurations of placing N particles in the n cells. Thus, a configuration is simply a list of numbers $(N_1, N_2, \ldots, N_n)$ where N_i is the number of particles in cell i. We assume that a cell can accommodate any number of particles, and we do not care for the precise location of the particles within cell i.

Even without doing the calculations, it is clear that the larger the number of cells n, the larger the number of configurations, hence also the larger the SMI. Take for example an extreme case: $n = 1$, i.e. only one cell, hence the number of configurations is one (all the particles are in the single "cell" of volume V), and the corresponding SMI is zero (we know where all the particles are). Clearly, for very large number of cells the SMI will be large too.

Thus, the SMI associated with the locations of the particles will change widely according to the precision we choose for the locations of the particles. However, whatever the choice of n is, when we let the gas expand from V to $2V$, the change of the SMI in this process will be $\log 2^N$, independently of n (provided we use the same cell size for both the initial and the finite states, i.e. $v = V/n = 2V/2n$).

We can conclude that the SMI of the system depends on the precision we choose to determine the locations of the particles, but the change in the SMI is independent of this choice.

What about the entropy change? The entropy of an ideal gas of N particles in a volume V and total energy E is determined by the variables (E, V, N) and is *independent* of the choice we might make with regards to the precision of the determination of the locations of all the particles. Note however, that the value of the entropy depends on the Planck constant h, which is a measure of the uncertainty in the determination of both the location and the velocity of the particles along each coordinate. This uncertainty is *built in* in the entropy and cannot be changed at will. It is one of the remarkable features of the entropy of a classical ideal gas which is a property of a macroscopic system, yet it depends on the quantum mechanical constant h.

The change in entropy for the expansion process, from V to $2V$, is $k_B N \ln 2$, independent of any precision we choose for the locations of the particles. This entropy change reflects the change in the SMI ($N \log 2 = N$), which is due to the loss of one bit per particle in the expansion process from V to $2V$.

2.7.2 *Does Entropy Depend on Our Knowledge of the Intermolecular Interactions?*

We have seen that the entropy of an ideal gas can be calculated given the energy E, the volume V, and the number of particles N. Does the entropy of the gas depend on whether we know, or do not know, the strength of the interactions among the particles?

For concreteness, suppose that I give you a box containing N atoms of argon in an isolated system of volume V. Does its entropy depend on the extent of the interactions between the particles?

Of course, the entropy *does* depend on the extent of interaction. However, it does not depend on whether we know the extent of these interactions. If you calculate the entropy of the system assuming that it as an ideal gas, you will get one value for the entropy; denote it by $S(ideal\ gas)$. This value will be different from the value you will calculate for the same system, but taking into account the intermolecular interactions among all the atoms. Let us denote the value of the entropy you will calculate by S(*real gas*). One can show that S(*real gas*) is smaller that S(*ideal gas*). Does this mean that the entropy of the system depends on the information you have on the system?

The answer is "No!" The entropy of the system you will calculate depends on whether you assume that the atoms interact, or do not (ideal gas). However, those entropies you calculate are not necessarily the true, or the absolute, entropy of the system.[11]

If you have a system with interactions, the true value is S(*real gas*). If you assume that it is an ideal gas, you will calculate a higher value of entropy S(*ideal gas*). But the latter value is the wrong value of the entropy.

The same is true if you have an ideal gas but you assume that there are interactions. In this case, the value you will calculate S(*real gas*) will be lower than the true value of the gas which is S(*ideal gas*).

Not only will the value of the entropy be different depending on the interactions between the particles, but if you do an experiment, say of expansion, you will get different values for the entropy changes. For concreteness, suppose that you have a system of one mole of argon, and you do the expansion experiment. If the system is an ideal gas, the entropy change is $\Delta S = R \ln 2$. This entropy change will be the same if the system is isolated, or at constant temperature. If the system is a real gas, the entropy change will have a different value from $R \ln 2$. The change in entropy will also be different if you do the experiment under constant energy (isolated system), or at constant temperature (isothermal system).

In the first experiment (isolated system) the entropy change will be due to three contributions. The first contribution will be due to the change in the locational distribution of the particles, i.e. $R \ln 2$. The second contribution will be due to the change in the interactions. The third will be due to change in the distribution of velocities, or equivalently, the change in the temperature.[12]

If on the other hand, the experiment is carried out at constant temperatures, then we have different contributions to the entropy change. One will be due to the expansion (i.e. the locational distribution) which is $R \ln 2$. The second will be due to the change in the extent of interactions between the atoms. There will be no change in the temperature in this case, but there will be a flow of heat into the system[13] which will also increase the entropy of the system.

Thus, we see that the true value of the entropy of the system, as well as the entropy change in the expansion process, does *not* depend on our knowledge about the extent of interactions. The calculated entropies will be dependent on our knowledge. However, if we have the wrong information on the extent of interactions, we will calculate the wrong entropy, as well as the wrong entropy change.

Finally, it should be noted that the above conclusion is not specific to entropy. It applies to any thermodynamic quantity. For instance, the total energy of the system depends on the extent of interactions. However, if we do not know the strength of interactions, we will calculate a wrong value of the total interactions. We use this example because many authors claim that entropy is a subjective quantity, and it depends on the information one has on the system.

The fact is that it is the calculated entropy that depends on the information we have, or assume, on the system, not the real, true entropy of the system.

2.7.3 Does Entropy Depend on the Knowledge of the Composition of the System?

Consider an ideal gas of $2N$ atoms of A. You are told that all the $2N$ atoms are indistinguishable. You calculate its entropy and find the values $S(2N\ of\ A)$. Next, you are told that the system contains two isotopes of A. For simplicity, assume that there are exactly N atoms of A and N atoms of A^*. The two isotopes have different masses, but for the moment we neglect this difference in mass. We only know that the two isotopes are distinguishable. You calculate the entropy of the system and you find out that its value is S(*N of A and N of A**), which is different from the value of S(*2N of A*). I have heard several times that people bring up this type of experiment to "prove" that entropy is a subjective quantity, i.e. that entropy depends on our knowledge about the system.

Does the entropy of the system depend on our knowledge of the isotopic composition of the system? This question is often stated as follows: Suppose we mix two ideal gases, one colored red, and the second, blue. In an experiment such as the one depicted in Figure 2.9a we observe a positive change in the entropy (the so-called "entropy of mixing"). What if someone who is color-blind does the same experiment? For the color-blind person, there is no change in the system, much as in the case of Figure 2.9b. Will the entropy change for such a person be zero?

This question becomes more subtle when we mix two isotopes of the same molecule, i.e. having the same color. In such an experiment, we are all "blind" to the isotopic composition of the system. An even more challenging question is the mixture of two enantiomers. This case will be discussed in Section 3.9.

Does the entropy of the gas, or the entropy change in the mixing of two different gases, depend on our knowledge of the isotopic composition?

The answer to this question is similar to the one given in the previous example (see Section 2.7.2). The entropy of the system *does* depend on the isotopic composition. The entropy we will calculate will be different, depending on our knowledge of the isotopic composition of the system. If we assume that the system is pure A, but in reality it is a mixture, and calculate its entropy we will get the wrong entropy. Likewise, if we assume it is a mixture, but in reality it is pure A, we will also calculate the wrong entropy.

To see this, let us do the following two experiments. In Figure 2.12a, we start with two boxes each having N molecules of

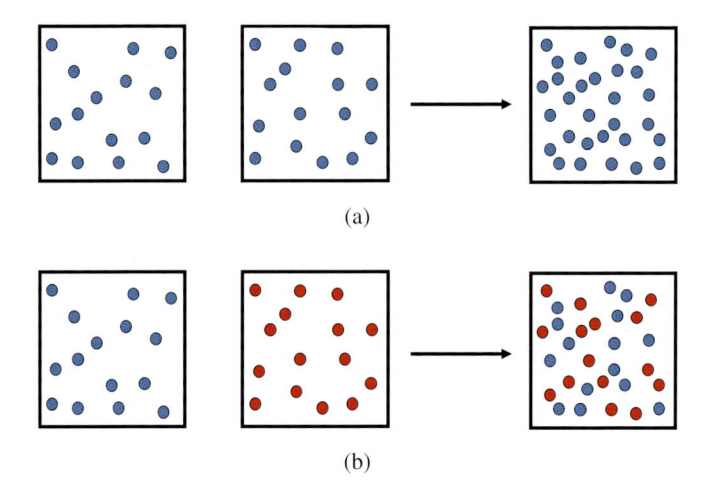

Fig. 2.12. (a) "Mixing" the same kind of molecules (pure assimilation), and (b) "pure" mixing of two different kinds of molecules.

pure A, in a volume V, and energy E. We assume that the systems are all ideal gases, i.e. there are no interactions among the particles. The entropy of each system on the left-hand side of Figure 2.12a is denoted by $S(N\ of\ A)$. We bring the two gases into the same box having the same volume V. The energy of the new system on the right-hand side of Figure 2.12a is twice the energy of each system on the left-hand side. What about the entropy? It is easy to calculate the entropy of the new system, which we denote by $S(2N\ of\ A) = 2S(N\ of\ A) - 2Nk_B \ln 2$. Thus, the entropy change in this process is $-2Nk_B \ln 2$. The reason for this lowering of the entropy is that initially, we had two systems each of N indistinguishable particles, and in the final state we have one system with $2N$ indistinguishable particles.[14] There is no change in the volume accessible to each particle, nor any change in the velocity distributions (or the temperature) of the gas (remember we are dealing with an ideal gas).

Now, we do almost the same experiments as in Figure 2.12a, but with two different isotopes A and A* (the volumes and energies being the same as before). We neglect the difference in the masses of the two isotopes. We calculate the entropy of each of the two systems on the left-hand side of Figure 2.12b, and we find that they are the same, i.e. $S(N \text{ of } A) = S(N \text{ of } A^*)$. We bring the two gases into one box of the same volume V and find that the entropy is simply the sum of the entropies of the two boxes on the left-hand side, i.e. $S(N \text{ of } A \text{ and } N \text{ of } A^*) = S(N \text{ of } A) + S(N \text{ of } A^*)$. The entropy does *not* change in this process. The reason is simple: the volume accessible to each particle and the distribution of velocities did not change in this process. In contrast to the previous experiment, the number of indistinguishable particles in this experiment did not change. On both sides of Figure 2.12b, we have N indistinguishable particles of one kind and another N indistinguishable particles of another kind of particles. In thermodynamics, this process is referred to as reversible mixing of two ideal gases.

Now suppose I give you one of the boxes on the right-hand side of either Figure 2.12a or 2.12b. You do not know if it is pure A, or a mixture of two isotopes A and A*. How would you calculate the entropy of the gas?

Clearly, your calculation of the entropy of the gas will be different if you assume that the gas is pure A, or a mixture of A and A*. This does not mean that the entropy of the gas is a subjective quantity depending on what you chose to assume for the composition of the system.

If the gas is a mixture of A and A*, and you do not know this, then you will calculate the wrong entropy of $S(2N \text{ of } A)$. If, on the other hand, if you assume that the gas is a mixture, but in fact it is pure A, you will again calculate the wrong entropy of $S(N \text{ of } A \text{ and } N \text{ of } A^*)$.

Pause and ponder

Suppose that A and A* are two enantiomers of say, alanine molecules (Figure 2.10). In this case A and A* have the same mass and the same energy levels. Repeat the same arguments as above for this case. Compare your answers with the discussion in Section 3.9.

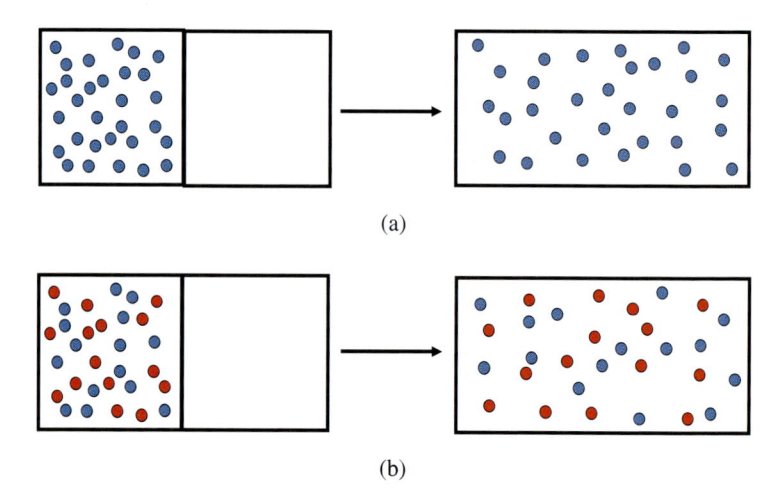

Fig. 2.13. (a) Expansion of pure A molecules, and (b) expansion of mixture of A and B molecules.

Finally, we note that whatever the value you will calculate for the entropy of the system of $2N$ atoms of A, if you carry out the expansion process in Figure 2.13 you will get the *same* change in entropy which is $2Nk_B \ln 2$. Here, in both of the processes in Figure 2.13, each particle has access to volume V in the initial state, but to volume $2V$ in the final state. The change in entropy is $2Nk_B \ln 2$ independent of the isotopic composition of the gases in the two experiments.

2.8 Entropy as a Measure of Freedom

In 1997 Nordholm proposed the following identifications:

$$\text{energy} \Leftrightarrow \text{wealth}$$

$$\text{entropy} \Leftrightarrow \text{freedom}$$

In this identification the Second Law is translated into the tendency of humans seeking for "maximized freedom." Although Nordholm admits that it is difficult to define precisely, let alone maximize, *freedom*, he believes that freedom is a "fair translation of the concept of 'entropy' from thermodynamics of inanimated matter to thermodynamics of human life."

What Nordholm has proposed is not an interpretation of entropy (the thermodynamic entropy), but a possible use of the SMI (which he refers to as entropy) in economics. This, of course, has nothing to do with entropy.

2.9 Entropy as a Measure of Possibilities

In 2000, Styer also suggested the interpretation of "entropy as freedom." He argued that as "freedom" means a range of possible actions, "entropy" means a range of possible microstates.[15]

The "possibility" interpretation of entropy has recently appeared in the book by Lemons (2013). Possibilities, as an alternative word for "multiplicity," is okay for an isolated system where all states are equally probable. In the epilogue of his book the author discusses the various interpretations of entropy. See also Section 2.2.4.

About the author's favorite interpretation, he writes:

> Entropy as possibility is my favorite short description of entropy because possibility is an apt word and, unlike uncertainty and missing information, has positive connotation. Thus, according to the second law of thermodynamics, an isolated thermodynamic system always evolves in the direction of the opening up of new possibilities.

This is true for isolated systems. In such a system, the terms "possibilities" and "multiplicity" are equivalent. It does not apply for all systems, e.g. systems at constant temperature, pressure and, composition. For instance, if you add a quantity of heat to a system, its temperature as well as its entropy will increase. The increase in entropy is not a result of an increase in the "possibilities," but is a result of the change in the distribution of the velocities of the particles. Thus, in general, entropy is not a measure of "possibilities," and the argument "because possibility is... has positive connotation," cannot be used in deciding which the right interpretation of entropy is.

Clearly, the "possibilities" interpretation of entropy cannot be used to understand why the net entropy change is positive in the example shown in Figure 2.7 (expansion of an ideal gas in a gravitational field).

The "possibilities" interpretation of entropy becomes misleading when the author explains the symbolic, beautiful image on the cover of his book:

> The image on the cover of the book almost perfectly captures the sense of entropy as possibility. As the constraints that inform a living organism dissolve, the entropy of the organism increases. As the flower dies, its seeds are scattered in the breeze. Yet, even in its death, new probabilities are sown.

Unfortunately, this description is both incorrect and misleading. Entropy is not "possibilities," and the entropy of a living organism does not increase. In fact, it is not even defined. Sadly, we must conclude that the beautiful image on the cover of the book is irrelevant to either entropy, or to the Second Law of Thermodynamics.

2.10 Entropy as a Measure of Irreversibility

In a glossary of Lemons' (2013) book, we find the following definition of entropy:

> Entropy: A measure of the irreversibility of the thermodynamic evolution of an isolated system.

This is a very unfortunate statement!

First, note that irreversibility means several things.[16] Second, "irreversibility" is a term used to describe a *process*, whereas entropy is a quantity assigned to a *state* of a system. Third, irreversibility is not a measurable quantity. Neither the value of the entropy of a system nor the change in entropy can be a measure of the irreversibility — certainly not the "irreversibility of the thermodynamic evolution of an isolated system."

Entropy, is a state function. This means that for any well-defined thermodynamic system, say (T, P, N), the entropy is also determined. Specifically, for an isolated system defined by the constant variables E, V, N, the entropy is fixed. The value of S has nothing to do with the "evolution" of the system, nor with any process that the system might, or might not, go through. Figure 2.14 shows two systems (E, V, N) and $(E, 2V, N)$. The latter has a higher entropy than the former. Does this mean it is also more "irreversible" than the former?

In Lemons' text, we find yet a different "definition." On page 9, the author states:

> ...the entropy difference between two states of an isolated system quantifies the irreversibility of a process connecting those two states.

Indeed, the entropy difference is by definition positive for a spontaneous process in an isolated system. However, reversibility or irreversibility of a process is not quantifiable.

In Figure 2.15 we show two processes of expansion. In 2.15a a mole of an ideal gas expands from V to $2V$. In 2.15b, one mole of gas expands from V to $3V$. The corresponding entropy changes in these two processes are $R \ln 2$ and $R \ln 3$, respectively. Clearly, the two processes are irreversible in the thermodynamic sense, i.e. that

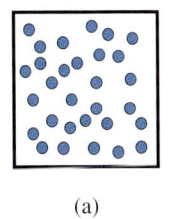

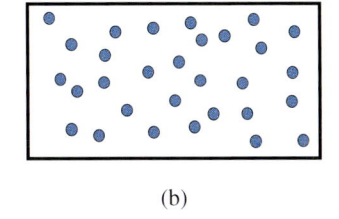

Fig. 2.14. (a) An (E, V, N) system, and (b) an $(E, 2V, N)$ system. The entropy of (b) is larger than that of (a). Which system is "more irreversible?"

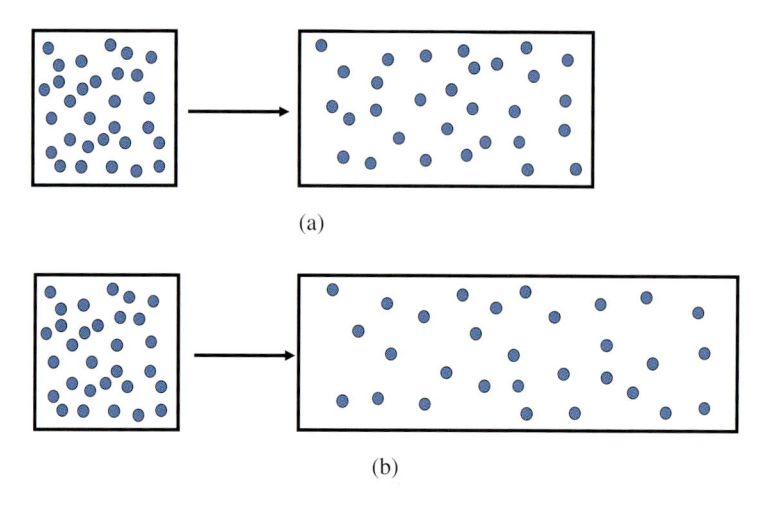

Fig. 2.15. Expansion of ideal gas (a) from V to $2V$, and (b) from V to $3V$.

$\Delta S > 0$. However, the second process, having a larger change in entropy, cannot be said to be more irreversible than the first.

In conclusion, "irreversibility" in an isolated system is not a quantifiable term. A process is either reversible or irreversible (depending of course on how we define reversibility of a process),[16] but neither entropy, nor entropy difference "qualifies the irreversibility of a process."

2.11 Entropy as "Heat Loss," "Thermal Energy Not Available To Work," "Unavailable Energy," Etc.

In most textbooks, as well as in dictionaries and encyclopedias, you find interpretations of entropy as a measure of the quality of the energy of the system: high grade or lower grade energy, etc. For instance, in Atkins (2007) we find the following statement[17]:

> U is a measure of the quantity of energy that a system possesses, S is a measure of the quality of that energy: Low entropy means high quality, high entropy means low quality.

Look again at Figure 2.14. On the right-hand side, the system has twice the entropy of the system on the left. This fact does not tell you anything about the "quality of the energy," the "availability of energy to do work," or any other description associating entropy with energy or work.

Entropy is a monotonic increasing function of energy (keeping V and N constants). It has nothing to do with the "quality" of that energy. Thus, Atkins' statement quoted above is meaningless, both quantitatively and qualitatively.

It is true that in some specific systems, say (T, V, N) or (T, P, N), the term $T\Delta S$ may be associated with "heat change," or the amount of work the system can do. However, this correct interpretation of $T\Delta S$ is due to the *product* of T and ΔS, not to ΔS alone. As I have posited earlier (see Note 1 of this chapter), and in more detail in Ben-Naim (2008), the reason for such a misinterpretation of ΔS is a result of the fact that the entropy carries the units of energy. This is the result of a historical accident. The temperature should have been measured in units of energy, and the entropy (as an SMI) should be either dimensionless or measured in units of bits. This will remove much of the confusion associated with entropy, as well as facilitate the acceptance of entropy as a special case of the SMI.

2.12 A Challenging Problem

Figure 2.16 shows a sequence of eight systems, all having the same fixed energy, volume, and total number of molecules. Each blue circle represents one mole of the *l* enantiomer, and each red circle represents one mole of the *d* enantiomers (see Figure 2.10). The systems differ only in the mole fractions of *l* and *d*.[18]

The method used to construct these systems is not important. One can think of starting with 10 moles of pure *l* molecules, system (a), or pure *d* molecules, system (h), then add a catalyst to convert some of the molecules from *l* to *d*, or from *d* to *l*.

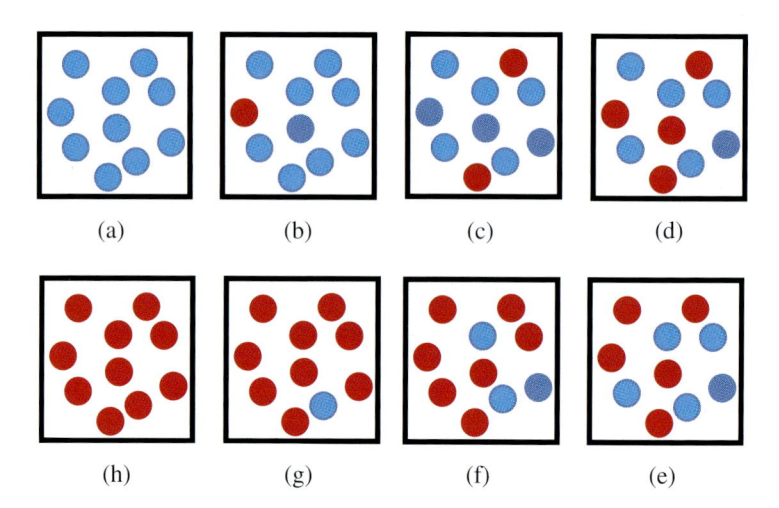

154 | Entropy: The Truth, the Whole Truth and Nothing But the Truth

Fig. 2.16. Eight isolated systems having different compositions of two enantiomers: l — blue circle, d — red circle.

Answer the following questions before consulting Note 19:

1. Suppose we start with pure l-molecules, system (a), and add a catalyst. Can you predict the final equilibrium state of the system, i.e. the mole fractions of l and d, at equilibrium?
2. What will be the change in entropy in the process described in the previous question?
3. Can you calculate the change in entropy of the system in going from (a) to (b) to ... (h)?
4. Can you rank the systems (a) to (h) according to
   (i) the extent of disorder (chaos, mixed-upness)
   (ii) the extent of energy spreading
   (iii) the extent of freedom of particles
   (iv) the extent of irreversibility of the system
   (v) the extent of information on the system
   (vi) the relative SMI in going from (a) to (h)
5. Which of the answers you gave to (i) to (vi) correlates best with the change in entropy?

2.13 Caveat on the Interpretation of Entropy as Uncertainty

There are many books where you can find an interpretation of entropy as *uncertainty*. For instance in Kafri and Kafri we find: "Entropy represents the uncertainty of a system."

As we noted in Section 1.4, this is true but not the whole truth. Colloquially, one can talk about the uncertainty about the color, the number of particles, the composition, the size, the weight, etc. All these are valid uncertainties, but have nothing to do with the SMI or with entropy. When one says that entropy is a measure of uncertainty, one must specify uncertainty with respect to what!

At this point there is a common error committed by many authors of popular science books. Looking at the Boltzmann equation, they conclude that it is the uncertainty about the state in which the system is. If we know the state in which the system is, then entropy of the system is zero. Such an interpretation is not valid, either to the SMI or to entropy.

Let us discuss briefly the uncertainty interpretation of entropy. You are given a system at equilibrium. You calculate the distribution of location and velocities. Then you calculate that distribution which maximizes the SMI. The entropy is related to the uncertainty associated with that distribution of ideal gas. If there are interactions, then you have to add the SMI associated with the interactions. In the particular case when we have an isolated system and we assume that all microstates are equally probable, we can say that the entropy is associated with this particular distribution. Note that this is true provided you *know the distribution*, or at least you assume that this is the distribution of the states.

Entropy, in general, is not related to the number of states as implied by the Boltzmann equation, but to the distribution of states as determined by the specific system.

If the system is not at equilibrium, then we do not know the distribution. One can guess the distribution, and calculate the SMI associated with that distribution, but again that is not the entropy; the entropy, by definition, is the SMI associated with the distribution which maximizes the SMI, which is the equilibrium distribution.

Clearly, saying that entropy measures the uncertainty about the system or about a living system or about the universe, is meaningless, unless one specifies uncertainty with respect to what? Once you do specify a distribution, you have a well-defined uncertainty which is measured by the SMI. For a thermodynamic system, once you have a well-defined distribution of microstates, then you also have a well-defined SMI of that thermodynamic system. It is only after you take the distribution which maximizes this SMI, i.e. at equilibrium, that you get the entropy.

2.14 Conclusion

We have seen that there are many invalid interpretations of entropy and to the best of my knowledge there is only *one* simple, intuitive, and proven interpretation based on the SMI.

There are of course, many descriptors of the state or changes of states of the system. The most common one is in terms of the extent of order or disorder. One can describe the state of the gas, relative to a liquid, and say that the gas is more, or less, ordered than the liquid. Such a description is in the eyes of the beholder. One does not, and one need not, prove the validity of such a descriptor.

On the other hand, when one claims that entropy is a measure of disorder (or spreading, freedom, or whatever), such a claim must be proven, and to the best of my knowledge, none of the common interpretations of entropy is proven to be correct. The interpretation of entropy as a special case of the SMI stands alone in a jungle of interpretations of entropy. It is a clear, meaningful, and informative interpretation. It is also very general as it applies to any well-defined system at equilibrium.

3

Applications and Misapplications of Entropy

3.1 Introduction

Entropy and the Second Law are extremely useful concepts in chemistry and physics. They are also useful in life sciences and cosmology. However, in the latter fields, entropy and the Second Law may be used only to study some aspects of biological systems, or some well-defined aspects of the universe — not life itself, neither the whole universe.

Even when we have a well-defined thermodynamic system, the way we use the entropy and the Second Law depends on the specific characterization of the system. Here is an example commonly featured in many popular science books:

When you boil an egg, it goes from a relatively *ordered* state to a *disordered* state, i.e. entropy increases. You never see a boiled egg, unboiling, i.e. entropy never decreases.

This example is usually given to demonstrate the one-directional change in entropy. Unfortunately, such an example is faulty for several reasons.

First, entropy is not always a measure of disorder (in fact, it is not even clear that a boiled egg is less ordered than an unboiled egg).

Second, it is not clear that entropy can be defined for a whole egg; either in its boiled or unboiled states. And if it could have been defined, it is far from clear that the entropy of the boiled egg is higher than the unboiled egg.

Finally, assuming for the sake of argument that the entropy is defined for the boiled and the unboiled states of the egg, the process of boiling is not governed by the Second Law, formulated in terms of entropy, but more likely, in terms of Gibbs energy.

The example above was taken from the literature. An egg is not a well-defined thermodynamic system, and the boiling of an egg is not a well-defined thermodynamic process. What I have said before is very qualitative. Let us now make a few more precise statements. Recall that the SMI can be interpreted as either the *uncertainty*, *unlikelihood*, or *measure* of *information* associated with a given probability distribution.

We have seen that entropy is a special case of the SMI. Therefore, entropy has exactly the same interpretation as the SMI when it applies to a specific set of probability distributions of a thermodynamic system at equilibrium. In other words, entropy can always be interpreted as uncertainty, unlikelihood, or a measure of information associated with the specific set of distributions at equilibrium.

This is true for any characterization of a thermodynamic system. It can be an isolated system (E, V, N constant), a closed system at constant temperature and pressure (T, P, N constant), etc. Therefore, entropy is useful to characterize the uncertainty, or the amount of information associated with the specific distributions at equilibrium.

While the meaning of entropy is the same for any well-defined thermodynamic system at equilibrium, the formulation of the Second Law is different for the different thermodynamic systems. Specifically, entropy does not necessarily increase for any spontaneous process.

When we remove a constraint in an isolated system, (say, remove a partition between two compartments or add a catalyst to a mixture of reagents), the system will proceed to a new equilibrium state having a higher entropy. On the other hand, when we remove a constraint in a system under constant T, P, N, the system will proceed to a new equilibrium state for which the Gibbs energy will be lower. In this case, the entropy change may be positive or negative, but the Gibbs energy change given by $\Delta G = P\Delta V + \Delta E - T\Delta S$, must be negative. Note also that the meaning of ΔS is the same in the T, P, N system as it is in the E, V, N system, but its role in the formulation of the Second Law is different in the two cases. In Table 3.1 we list a few common cases of thermodynamic systems. There are other systems, such as open systems or systems under the influence of an external field, for which the meaning of entropy is the same, but its role in the formulation of the Second Law is different. See Ben-Naim (2012, 2015a).

In the following sections of this chapter, I will present a few examples of the uses of entropy and the Second Law of Thermodynamics. Of course, this is a very small sample of examples for which entropy and the Second Law (in its various formulations) are applied. I have chosen only a few examples with which I am familiar with to demonstrate the usefulness of the concept of entropy and the Second Law.

Table 3.1 Various Thermodynamic Systems, their Thermodynamic Potentials, and the Corresponding Formulation of the Second Law

Thermodynamic Characteristic	Thermodynamic Potential	Second Law
E, V, N	Entropy S	$\Delta S \geq 0$
T, V, N	Helmholtz energy A	$\Delta A \leq 0$
T, P, N	Gibbs energy G	$\Delta G \leq 0$

In the succeeding sections of this chapter I will very briefly mention a few misuses of entropy and the Second Law. Some of these misuses can well be termed as abuses, as there is a thin line between misuse and abuse, and the demarcation line that separates them is not always well defined.

3.2 Residual Entropy

One of the most interesting and useful aspect of the entropy itself (i.e. not involving the Second Law) is the so-called *residual entropy*.

Recall that Clausius' definition of entropy, as well as most of the applications of entropy in thermodynamics, involves the difference between the entropy values of the system at two different states. On the other hand, Boltzmann's definition, as well as the definition based on the SMI, provides "absolute" values of the entropy.

I enclosed "absolute" in inverted commas because up to this point, we cannot verify that the calculated theoretical values obtained from the entropy function are the "true" values. How can we compare the calculated theoretical values of entropy with experimental results? Here, the Third Law of Thermodynamics comes to our aid. The Third Law basically states that the entropy of a system tends to zero when we approach the absolute zero temperature ($T = 0K$). Whenever this is true, one could in principle calculate from experimental data (on heat capacity, heat of melting, heat of evaporation, etc.) the difference between the entropy of the actual system and the entropy of the system at $T = 0K$. If the entropy of the system at $T = 0K$ is indeed zero, then the difference $S(T) - S(T = 0)$ will give us the "absolute" value of $S(T)$.

Indeed, when one calculates the entropies of some gases based on experimental data and compare them with the entropies obtained by the theoretical entropy function, one gets a very good agreement between the two results. In such cases, we can remove the inverted commas and refer to the entropy as the absolute entropy of the gas.

Table 3.2 Calculated and Measured Entropies of Gases

		Entropy (E.U.)	
Substance	Temperature	Theoretical	Experimental
A	B.P.	30.87	30.85
O_2	B.P.	40.68	40.70
N_2	B.P.	36.42	36.53
Cl_2	B.P.	51.55	51.56
HCl	B.P.	41.45	41.3
HD	298.1K	34.35	34.45
CH_4	B.P.	36.61	36.53
C_2H_4	B.P.	47.35	47.36
NH_3		44.1	44.1
CO_2		47.5	47.5
CH_3Br		58.0	57.9
CH_3Cl	298K	55.9	55.9

B.P. is the boiling temperature. E.U. are the entropy units of cal/mol. K.

Table 3.2 shows values of the entropy of a few gases calculated from experimental data, and from the theoretical equation for the entropy.[1]

The agreement between the theoretical and the experimental values is, in fact, the proof for the success of the Boltzmann's entropy, as well as the entropy, based on the SMI. Without this agreement we could not know the absolute value of the entropy for any substance.[2]

There are however, many cases where there is no agreement between the theoretical and the experimental values. We denote by S_{exp} the experimental value, and by S_{theor} the theoretical value of the entropy. Table 3.3 shows a few examples where there is a discrepancy between S_{exp} and S_{theor}. We define the *residual entropy* by the difference between the experimental and the theoretical values: $\Delta S = S_{theor.} - S_{exp}$.

Table 3.3 Calculated and Measured Entropies of Gases

Substance	Temperature	Theoretical	Experimental	Residual Entropy
CO	B.P.	38.32	37.2	1.22
NO	B.P.	43.75	43.03	0.72
N_2O	B.P.	48.50	47.36	1.14
H_2O	298K	45.10	44.29	0.81
D_2O	298K	46.66	45.89	0.77
CH_3D	B.P.	39.49	36.72	2.77

E.U. are entropy units; see Table 3.2.

There are many reasons for discrepancies between the theoretical and experimental values of the entropy. There could be inaccuracies in the values of the heat capacities, or the heat of melting, or the heat of boiling. [For details, see Wilks (1961).] However, there are some discrepancies which can well be explained by the so-called *configurational degeneracy* of the crystals at very low temperatures.

Consider the following simplest examples: linear molecules such as HCl, HBr, or HI. When the crystals of these molecules are cooled down to a very low temperature, there is only one arrangement of all the molecules in the crystal for which the crystal's energy is minimal. The reason is that if you change the orientation of the molecule, say from HCl to ClH, the energy of the crystal will change considerably. Therefore, as the temperature is lowered, the molecules will tend to orient themselves in such a way that the total energy of the system will be the lowest. In such cases it is believed that there is only one configuration of the molecules at 0K, hence $S(0) = k_B \ln 1 = 0$.

On the other hand, when you cool crystals of molecules such as CO, NO, and NNO, there is not much difference in the interaction energy between the molecules having either the NO or ON configurations. Figure 3.1 shows a possible configuration of a crystal NO. If

NO NO ON NO NO NO NO ON NO NO
ON NO NO ON NO NO ON ON NO NO
ON ON NO NO ON ON NO ON ON NO
ON ON NO ON ON ON ON NO ON ON
ON NO NO ON NO NO ON ON NO NO

Fig. 3.1. A possible configuration of a crystal of NO. It is assumed that the energy of the crystal is almost independent of the orientations of the molecules: NO or ON.

this assumption is correct, then we can calculate the total number of configurations in a crystal of one mole of NO. Assuming that each molecule can be in either one of the two orientations (CO or OC, NO, or ON, NNO or ONN), then there are altogether 2^N possible configurations. Assuming that the energy of all these configurations are nearly equal, we can calculate the Boltzmann entropy for this configurational degeneracy, which is $k_B \ln 2^N \approx 1.38 \, \text{cal} \, \text{mol}^{-1} \text{K}^{-1}$.

This is very close to the residual entropy of the molecules such as CO and NNO (and to some extent NO) as shown in Table 3.3.

Another interesting example is the case of methane (CH_4), methyl chloride (CH_3Cl), and methyl bromide (CH_3Br). In all of these molecules the residual entropy is nearly zero (see Table 3.2). On the other hand, if we take the molecule CH_3D (i.e. methane in which one hydrogen is replaced by deuterium), we find a considerable residual entropy (see Table 3.3).

Again, the approximate explanation for this residual entropy is in terms of configurational degeneracy. Suppose that in the crystal of pure methane the arrangement of the molecules is as shown in Figure 3.2. Clearly, if you rotate the molecule about any of the CH axes by an angle of $2\pi/3$, you will get exactly the same configuration, hence the same energy of the crystal. In this case, there is no configurational degeneracy. Now consider the molecule CH_3Cl (see Figure 3.3). If you rotate the molecule about the axis C–Cl by $2\pi/3$, you will get exactly the same configuration. However, if you rotate about the C–H axis by an angle of $2\pi/3$, you will get a new configuration which might

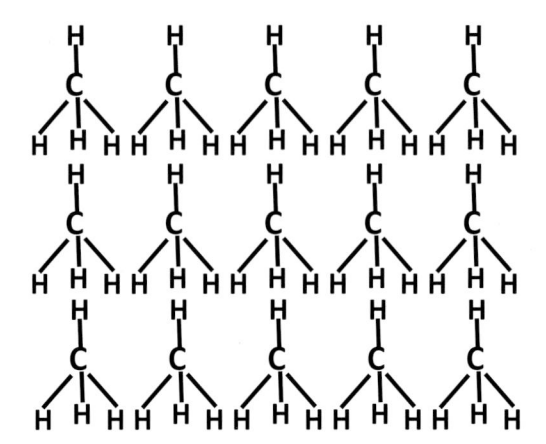

164 | Entropy: The Truth, the Whole Truth and Nothing But the Truth

Fig. 3.2. A possible arrangement of methane molecules in a crystal.

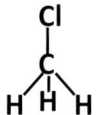

Fig. 3.3. Methyl chloride.

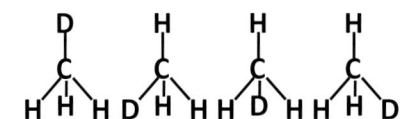

Fig. 3.4. Four different configurations of deuterated methane.

have a very different interaction energy with neighboring molecules. Therefore in this case, when you cool down the crystal of CH_3Cl, all the molecules will be oriented in such a way that the total energy of the crystal will be minimal. In this case there will be only one configuration, hence the entropy at 0K will be $S(0) = 0$.

The situation is much different for the case of CH_3D. In this case there are four configurations (see Figure 3.4) which are different (unlike the case of methane). On the other hand, these four configurations, unlike CH_3Cl, have very little difference in the interaction energies among the molecules. Therefore, assuming that each molecule of CH_3D can be in any one of the four configurations, then for N molecules we will have 4^N configurations. Assuming (approximately) that all these configurations have the same energy (therefore, equally probable) we arrive at the estimate of the residual entropy of CH_3D which is, for one mole of CH_3D,

$$k_B \ln 4^N \approx 2.75 \, \text{cal mol}^{-1} \text{K}^{-1}$$

which is very close to the value shown in Table 3.3.

We can conclude that if there are many configurations which have nearly the same energy, the system will have a residual entropy which may be calculated from Boltzmann's equation $S = k_B \ln \Omega$.

3.2.1 *Residual Entropy of Ice*

We present here the calculation, as well as the interpretation, of the residual entropy of ice. This is an important example demonstrating how a simple statistical calculation led to the understanding of the concept of the residual entropy of ice, and indirectly contributed to the understanding of entropy in general. This calculation was first published by Linus Pauling in 1935.

The residual entropy of ice is a measure of the number of configurations, or the number of arrangements of the hydrogen atoms on the fixed lattice of ice. Here we follow Pauling's procedure of counting the number of arrangements of the hydrogen atoms in a crystal of ordinary ice. It is remarkable that this number, which is a microscopic property of ice, can also be measured experimentally in the macroscopic world.

The structure of ordinary ice was determined by Bragg in 1922 using X-ray crystallography. This "structure" refers only to the arrangements of the oxygen atoms as shown in Figure 3.5. Each atom

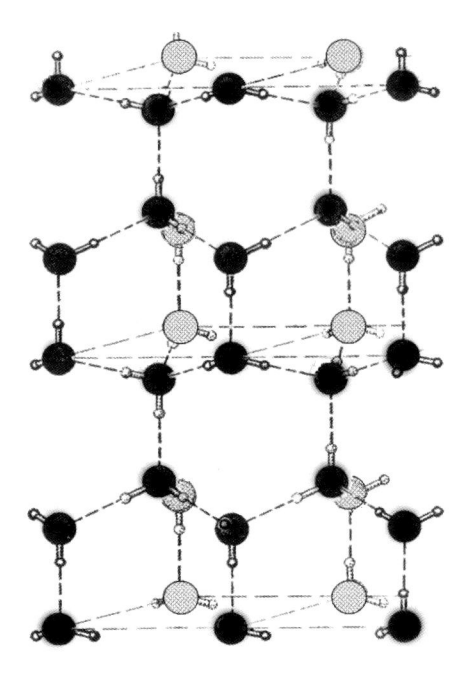

Fig. 3.5. The structure of ice.

of oxygen is surrounded by four atoms of oxygen situated at the vertices of a regular tetrahedron. Nothing was known on the "structure" of the hydrogen atoms in ice. It was known that the H_2O molecule retains its entity as a molecule in both the liquid and the solid phases.

Pauling asked himself the following question: In how many ways can one place $2N$ hydrogen atoms on the crystal structure of ice formed by N oxygen atoms?

Remember, we have N molecules of water. This means that there are N atoms of oxygen and $2N$ hydrogen atoms. Remember also that if you have N oxygen atoms you also have $2N$, O–O bonds. If you were sitting on one oxygen atom, you will observe from this vantage point that each atom of oxygen is connected to four neighboring oxygen atoms. You might have concluded that there should be $4N$, O–O bonds formed by N oxygen atoms. However, this is an over-counting. If you take all the N oxygen atoms and assign to each atom four O–O lines, then you have counted each O–O bond twice. One, when viewed from the first oxygen, say the one on the right of O–O, and one when viewed from the left oxygen. Therefore, the 4N count should be divided by 2 to get the correct number of O–O lines, or O–O bonds.

The question can be stated as follows: In how many ways can we arrange the 2N hydrogen atoms on the 2N O–O lines in such a way that the two *ice rules* are satisfied? The ice rules are (see Figure 3.6):

1. On each O–O bond there must be one and only one hydrogen atom.
2. Each oxygen atom must have only two nearby hydrogen atoms. This means that each water molecule retains its entity as composed of one oxygen and two hydrogen atoms.

The ice structure extends in all the three dimensions. However, for our counting process we shall use the two-dimensional diagram of ice (see Figure 3.7).

We now carry out the calculation of the number of arrangements in two steps.

First, suppose that we have only the first ice rule (1). We need to place the 2N hydrogen atoms on the 2N, O–O bonds. On each of

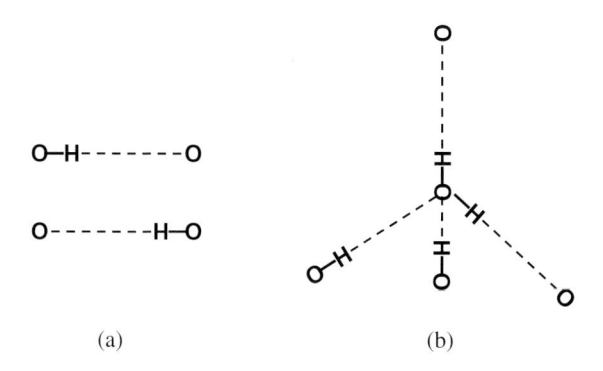

Fig. 3.6. The two ice rules: (a) each O–O bond has one H, and (b) each oxygen atom has two nearby hydrogen atoms.

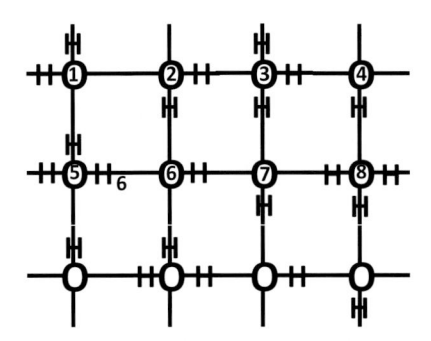

Fig. 3.7. Two-dimensional diagram of ice.

the O–O bonds there are two locations, one which is near the right oxygen, and the second near the left oxygen (see Figure 3.6a). In how many ways can we do this?

This problem is relatively easy to solve. Each hydrogen atom can be placed in either one of the two "sides" of the O–O bond. Therefore, there are altogether $2 \times 2 \times 2 \times 2 \cdots 2$, i.e. multiply 2 by itself $2N$ times. This number is 2^{2N}.

All these 2^{2N} arrangements satisfy the first ice rule (1). One out of many arrangements is shown in the two-dimensional diagram (see Figure 3.7).

Take note that each O–O line received only one H, as required by the first ice rule. As you can see, the second ice rule (2) is not always satisfied. Some oxygen atoms have more than two nearby hydrogen atoms, and some have less.

Second, we need to satisfy also the second ice rule (2). This part of the problem is a little trickier but not too difficult. It also involves some approximations.

Suppose we have made all the 2^{2N} arrangements that satisfy the first ice rule. We want to "extract" or to "choose" from all of these arrangements, only those that satisfy also the second ice rule. In Figure 3.7 we have one example of an arrangement of hydrogen atoms on the lattice of two-dimensional ice. We can see that the oxygen atoms (1) and (2) have two, nearby hydrogen atoms. These are "good"

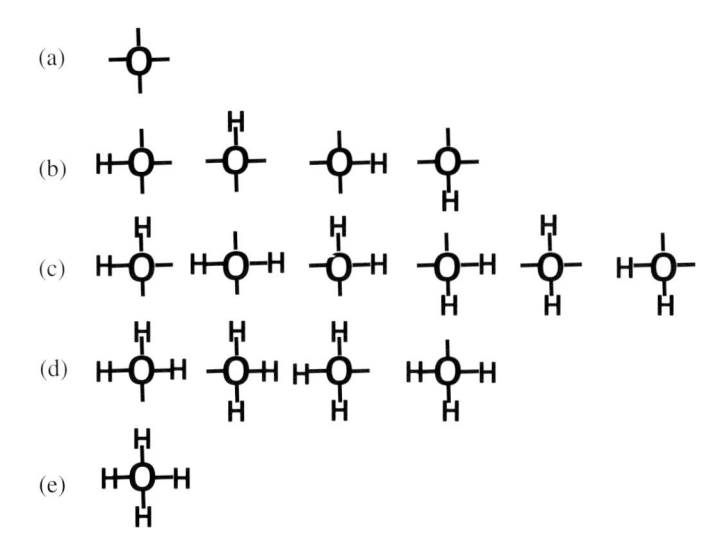

Applications and Misapplications of Entropy | 169

(a) [figure]
(b) [figure]
(c) [figure]
(d) [figure]
(e) [figure]

Fig. 3.8. All 16 possible arrangements around one oxygen.

configurations. However, the configuration about oxygen atoms (3) and (4) are not "good."

Clearly, we cannot go over all of the 2^{2N} arrangements to select the ones we are interested in. That is the point where human ingenuity comes in handy.

Let us follow Pauling's argument on how to select those arrangements, which satisfy both of the two ice rules out of a huge number of 2^{2N} arrangements. Consider all the possible configurations or arrangements of the hydrogen atoms viewed from the vantage point of sitting on one oxygen atom. Here are all the possibilities as seen from randomly selected oxygen atoms (see Figure 3.8).

As we can see, there are five classes of possible arrangements. Let us name them as follows:

Class zero — those configurations for which there are no nearby hydrogen atoms
Class one — those configurations for which there is only one nearby hydrogen atom

Class two — those configurations for which there are two nearby hydrogen atoms

Class three — those configurations for which there are three nearby hydrogen atoms

Class four — those configurations for which there are four nearby hydrogen atoms

So far we have classified all the possible configurations into five different classes according to the number of hydrogen atoms that are close to one base atom of oxygen.

Next, we have to count how many different configurations exist in each class. Remember that each class is characterized by the number of nearby hydrogen atoms. It is easy to count how many different configurations exist in each specific class. These are as follows:

Class zero has only one configuration.
Class one has four different configurations.
Class two has six different configurations.
Class three has four different configurations.
Class four has only one configuration.

All of these configurations are shown in Figure 3.8. Altogether, there are 16 different arrangements around a single oxygen, i.e.

$$1 + 4 + 6 + 4 + 1 = 16$$

Clearly, most of the configurations shown in Figure 3.8 do not conform to the second ice rule. Recall that the second ice rule requires that each oxygen have exactly two nearby H atoms. (see Figure 3.6b). It is immediately clear that only the six arrangements belonging to class two are the ones that satisfy the second ice rule.

So far we did a very simple counting. First, we counted all possible arrangements of $2N$ hydrogen atoms on the $2N$, O–O bonds. Next, we counted all the configurations of hydrogen atoms as seen from one oxygen atom. We found that only six out of sixteen possible arrangements also satisfy the second ice rule. Therefore, the fraction of valid configurations that satisfy both of the ice rules is 6/16. This is the fraction of the valid configurations about each oxygen atom.

Here comes the leap of the ingenuity of Pauling's thought. It involves an approximation, but a very plausible one.

Suppose you threw 1,000 dice on the table. You will find several different faces showing one, two, three, or more dots. Imagine that you pick up a die, and ask yourself what the chance (or the probability) is that the die you have picked up shows the number 4. Since there are six possible outcomes, the fraction of outcomes that are 4 is simply 1/6. In other words, we say that the probability of obtaining a specific result, say number 4, is 1/6. This means that if you repeatedly do the same experiment many times, say 1,000 throws, you will find that one in each six draws will be the number 4, that is, 1000/6 of the throws will show the number 4.

Similarly, we look at all the oxygen atoms having all the possible arrangements of hydrogens as shown in Figure 3.8 and ask: What is the chance that a randomly picked-up oxygen atom will have the arrangement as in class two? The probability of this event is the fraction 6/16. This is similar, but not exactly the same as, the calculation we have done for the dice (it involves the assumption of independence of all the 16 arrangements in Figure 3.8).

The second assumption is that the "events" about all the oxygen atoms are independent, therefore the probability (or the fraction) of all the N oxygen atoms to be found in a configuration belonging to the class two is the product of the probabilities of each atom to belong to class two in Figure 3.8, which is

$$\frac{6}{16} \times \frac{6}{16} \times \frac{6}{16} \times \cdots (N \; times) = \left(\frac{6}{16}\right)^N.$$

Finally, Pauling's argument is as follows: First, take the total number of arrangements that fulfill the first ice rule: 2^{2N}. This is a huge number for N of the order of 10^{23}. Then, assume that the fraction of all these 2^{2N} configurations which also fulfill the second ice rule is $(\frac{6}{16})^N$. This is a very small number. Multiplying these two numbers, he got the number of valid arrangements, i.e. arrangements that simultaneously fulfill the two ice rules, which is

$$\left(\frac{6}{16}\right)^N \times 2^{2N} = \left(\frac{6}{16} \times 4\right)^N = \left(\frac{24}{16}\right)^N = \left(\frac{3}{2}\right)^N = (1.5)^N$$

This is the total number of configurations (or different arrangements) of $2N$ hydrogen atoms on $2N$, O–O bonds, such that the two ice rules are fulfilled.

What is so remarkable is that this simple calculation, notwithstanding its approximation, is in excellent agreement with the residual entropy of ice. The entropy in this case is simply related to the logarithm of the total number of arrangements. For one mole of ice this is

$$S = k_B N \ln(1.5) = 0.805 \text{ cal/mol K}$$

Here, we have arrived at the results 1.5^N by purely theoretical calculation. The same quantity, called the residual entropy of ice, may be obtained experimentally, by heating pure ice from very low temperatures, up to room temperature. The experimental value is 0.81 cal/mol K. For more details on the calculation of the experimental and theoretical values of the entropy of water, see Rushbrooke (1949).

The agreement between the two, is again a victory of the validity of the theoretical values of the absolute entropies. In the case of water, it also explains that in ice, the hydrogen atoms are not at fixed lattice points but they can easily "jump" from one site to the other.

It should be stressed, however, that all the calculations based both on theoretical and on experiments are approximate. The relationship between the entropy and the number of states is strictly valid for a system for which all microstates (here, the number of configurations) have the same energy.

3.3 Application of the Second Law for Processes in an Isolated System

As we noted in Section 1.6 and in other sections in this book, a strictly isolated system is an idealized system. In practice, we never work or perform experiments on an isolated system. Nevertheless, such systems are extremely useful as the basis on which the whole edifice of statistical mechanics has been erected. The Second Law for such a system states that if we remove a constraint in an isolated system, the system will move to a new equilibrium state having higher entropy. In many cases, we can measure the entropy change for such a spontaneous process and compare the result with the theoretical calculated entropy difference. Thus, for the expansion of one mole of an ideal gas from a volume V to a volume $2V$, the entropy change is $R \ln 2$. This result can be obtained both experimentally, as well as theoretically. Again, such an agreement is a witness to the success of the theoretical function of the entropy. The interpretation of this result in terms of the SMI is simple. Each particle was initially confined to a volume V. In the final state we have "lost information," now we do not know whether the particle is in the right, or the left compartment. This means we lost one bit of information per particle. The change in the SMI for this process is simply $N \log_2 2 = N$, i.e. we lost N bits in the process. Similarly, in the process of mixing two ideal gases, as in Figure 1.2b, we have N particles which were initially confined to the left compartment, and another N particles which were confined to the right compartment. After we remove the partition between the two compartments, the loss of information is $2N \log_2 2 = 2N$ bits. The corresponding change in entropy is $2R \ln 2$. The interpretation of this result in terms of the SMI is exactly the same as in the case of the expansion. We lose one bit for each atom in the process. Since we have $2N$ atoms, each "expands" its accessible volume from V to $2V$; the total change in the SMI is $2N$.[3]

3.4 Applications of Entropy and the Second Law for a Constant Temperature and Constant Pressure System

The thermodynamic systems which are most often used in laboratory experiments are those which are closed (i.e. not exchanging materials with the surroundings), at constant temperature (kept by a thermostat), and at constant pressure (normally at 1 atm pressure, but this can be varied). We say we have a system characterized by T, P, N, where N stands for all atoms of the system being constant. Note that molecules can react to form new molecules, say two molecules of hydrogen (H_2) and one molecule of oxygen (O_2) can form two molecules of water. Such a process does not involve any change in the total number of atoms of hydrogen and oxygen in the system.

In such a system the meaning of the entropy is the same as in an isolated system. However, the role of entropy in connection with the Second Law is very different. In this system the Second Law does not state that entropy must increase upon a removal of a constraint. Instead, one defines a new function referred to as the Gibbs energy (sometimes also called the free energy, or Gibbs free energy) which is defined by $G = E + PV - TS$. Here, E is the energy of the system, P the pressure, V the volume, T the temperature, and S the entropy of the system.

In a T, P, N system, the Second Law states that in any spontaneous process occurring following the removal of a constraint, the Gibbs energy change will be negative. Since the process is at constant T and P we can write $\Delta G = \Delta E + P\Delta V - T\Delta S$.

The entropy change can be either positive or negative in such a process. The Second Law does not say anything about the magnitude or the sign of the entropy change.

In some of the following examples we shall ignore the term $P\Delta V$ and write approximately: $\Delta G \approx \Delta E - T\Delta S$. There are processes in the gaseous phase where $P\Delta V$ is not negligible and must be taken into account. However, in many examples in a condensed phase, we can ignore the term $P\Delta V$.

Thus, what determines the value of the Gibbs energy change is the change in the energy, as well as the change in the entropy.

In many popular science books you will find a general statement about the tendency of a spontaneous process to go from order to disorder — meaning positive change in entropy. Such statements are doubly erroneous; first, because not in any spontaneous process does the entropy increase, and second, the entropy is not always a measure of the extent of disorder in the system.

We present a few examples of spontaneous processes in a (T, P, N) system in which the change in entropy can be either positive or negative. The Gibbs energy change, however, must be negative as required by the Second Law.

3.4.1 *Formation of Hydrogen Molecules from Hydrogen Atoms*

The simplest chemical reaction is the formation of a hydrogen molecule (H_2) from two atoms of hydrogen. This reaction is written as

$$2H \rightleftharpoons H_2$$

Suppose we start with a dilute gas of atomic hydrogen at some specific temperature, pressure, and with one mole of H atoms. We assume that at this state the atoms do not react to form molecules. Let us imagine that there is an inhibitor that precludes the association of two atoms.

At some point we remove the inhibitor (or equivalently add a catalyst to the system), and we find that the system proceeds to a new equilibrium state. This is similar to the removal of a partition between two compartments. If we keep the temperature and the volume of the system constant, the spontaneous process which occurs is not governed by the entropy formulation of the Second Law, but by the free energy formulation (strictly here, the thermodynamic potential is the Helmholtz energy, to which we refer to, for simplicity, as the free energy).

In fact, in this particular reaction most of the hydrogen atoms will form hydrogen molecules. The change in entropy of the system would be negative. We can calculate the change in the entropy, and we find that the major contribution to the change in entropy is due to the loss of the translational entropy of the hydrogen atoms (there are small changes due to rotations and vibrations of hydrogen molecules which we ignore here).

The interpretation of this entropy change is very simple. In the initial state each atom of hydrogen could access the entire volume V of the system. In the final equilibrium state almost all the atoms form molecules. Now, we have about half a mole of molecules having access to the entire volume V. Looking at one *molecule* of hydrogen, we can say that one atom can wander and access the entire volume V. However, the second atom, attached to the first, is not free to access the entire volume V. Instead, because of its attachment to the first atom, it has a *reduced* volume of accessibility. At each point the first atom (1) is, there is a small spherical shell of volume v around it, which is accessible to the second atom (2). This is illustrated in a two-dimensional case in Figure 3.9. Thus, for half of the atoms, the total volume V is accessible. For another half, only the small region of volume v is accessible. This is a huge change in the accessible volume available to all atoms. Corresponding to this change in volume, there is a negative change in entropy. The interpretation of this negative change in entropy is simple. We initially know that each molecule could be at any point in the entire volume V. In the final equilibrium state we gain some information. If we know where one atom is, say at R_1 (within the entire system of volume V), then we also know that

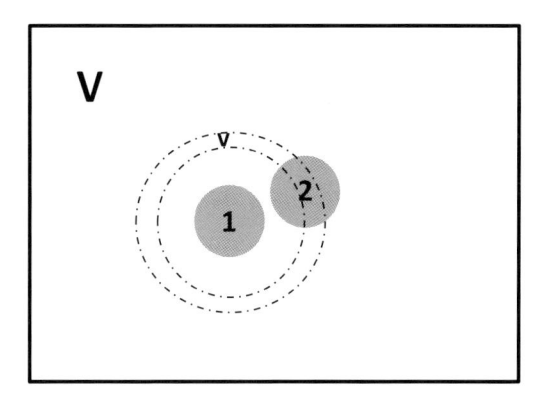

Fig. 3.9. A spherical ring of volume v about one atom (1), which is accessible to the second atom (2). Here the illustration is in two dimensions.

a second atom must be located at a small region around R_1. (Again, we note that there are other contributions to the entropy due to the rotation and vibration of the hydrogen atoms which are relatively small, and we ignore them in this discussion.)

Clearly, if one uses the Second Law in its entropy formulation, i.e. "entropy always increases in a spontaneous process," one would be puzzled to find that in this particular process the entropy of the system has decreased.

The Second Law, when applied to this system (i.e. T, V, N constant), states that the free energy (more precisely the Helmholtz energy) $A = E - TS$ will decrease in this process, i.e. $\Delta A = \Delta E - T\Delta S < 0$. When $\Delta S < 0$, $-T\Delta S$ is positive. Therefore, in order to obtain a net negative change in the free energy ($\Delta A < 0$), the change in energy ΔE must be negative, and in absolute value, it must be larger than $|T\Delta S|$, i.e. we must have $|\Delta E| > |-T\Delta S|$. Furthermore, the free energy will reach a minimum at the new equilibrium state. This minimum also allows us to calculate the *equilibrium constant* for this reaction. Basically, this means that we can calculate the fraction of atoms which form molecules, or in general the fraction of molecules of each kind at the final equilibrium state.

We can conclude by saying that in this spontaneous process there are two competing effects. The entropy change is negative, hence $-T\Delta S$ is positive. Therefore, this change in entropy contributes positively to ΔA. However, because of the strong chemical bond between the two atoms of hydrogen, there is a large negative change in energy of the system, $\Delta E < 0$. This change in energy must be large enough to more than compensate for the positive contribution of $-T\Delta S$.

As we have noted earlier, the "driving force," i.e. the reason for going from the initial to the final equilibrium state is probabilistic. The thermodynamic "driving force" in this case is the Helmholtz energy (see Sections 1.6.3 and 1.6.4).

3.4.2 *Protein-Protein Association*

In the previous example we saw the two competing effects of the energy and entropy to determine the free energy change in the process of dimerization of hydrogen atoms. In that example we had a clear-cut case where we know the origin of the negative entropy change on one hand, and the negative energy change, on the other hand.

Here we discuss a very similar process: dimerization of two globular proteins in an aqueous environment. We start with a system of one mole of monomeric protein at a given temperature T, pressure P, and N_W water molecules. We remove an inhibitor (or add a catalyst), and we find that a fraction of the protein molecules will form stable dimers (there are also cases of formation of tetramers and higher aggregates; the problem is the same in all of these cases). Figure 3.10 shows a schematic description of such a process. Since the system is at constant temperature and pressure, the relevant thermodynamic

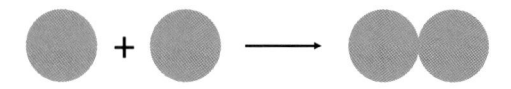

Fig. 3.10. Schematic process of dimerization of proteins.

potential function is the free energy (strictly, it is the Gibbs energy $G = E + PV - TS$, but we ignore the PV term).

As in the case of dimerization of hydrogen atoms to form "dimers," here some of the monomeric protein molecules form dimers. The main contribution to the entropy change in this process is also the same as in the case of hydrogen atoms, i.e. some of the monomers lose their translational entropy (i.e. initially they can access the entire volume V, whereas in a dimer they are confined to a small region around the other monomer as described in Section 3.4.1).

The negative entropy change is reasonably well understood (not fully understood because there are some changes in the water which also contribute to the entropy change. However, the main contribution to the change in entropy is the same as in the case of the formation of hydrogen molecules).

Remember, in this system the entropy is only part of the free energy. We found that $-T\Delta S$ is positive, and we know that the process is spontaneous. Therefore the free energy change in this process must be negative.

Here is the big problem. In the case of hydrogen atoms we know that the energy change in the process is large and negative. It is large enough to overcompensate for the positive quantity of $-T\Delta S$. The reason is the strength of the chemical bond between the two hydrogen atoms.

What is the source of energy change in the process of protein-protein dimerization which is so large to overcompensate for the negative entropy change?

This is a real mystery. It is considered to be one of the "big unknowns in science."[4]

We know that there are no covalent bonds between the monomers in the dimers (or in any other aggregate). We also know, or at least we believe, that the water molecule contributes significantly to the binding free energy. So the question is, What is the origin of the large negative free change for the process of dimerization of proteins? To put this question metaphorically, we ask, What is the "glue" that keeps the two partners bound together, despite the entropy effect which wants the monomer to stay single?[5]

The answer to this question is controversial. I will present here my personal view on this question. More may be found in Ben-Naim (2010, 2015a, and 2016a).

Globular proteins have many hydrophilic groups on their surface. These groups, such as hydroxyl (OH), carbonyl (CO), and amine (NH) can form hydrogen bonds with water molecules.

When two globular proteins come to a close distance, water molecules can form hydrogen-bond bridges between hydrophilic groups on one monomer, and hydrophilic groups on the other monomers, as shown in Figure 3.11. It has been estimated that such hydrogen-bond bridges can contribute significantly to the free energy change in this process. The large negative Gibbs energy change is mainly due to the large negative energy change, while the contribution of the entropy change, i.e. $-T\Delta S$ is positive.

Thus, we see that although there is no direct "glue" due to a chemical bond between the two protein monomers, there is an indirect "glue" which is due to a solvent-induced interaction between the two monomers.

To conclude, we can say that there are two contributions to the free energy change of the process of dimerization of proteins. The entropy change is negative, and this change contributes positively

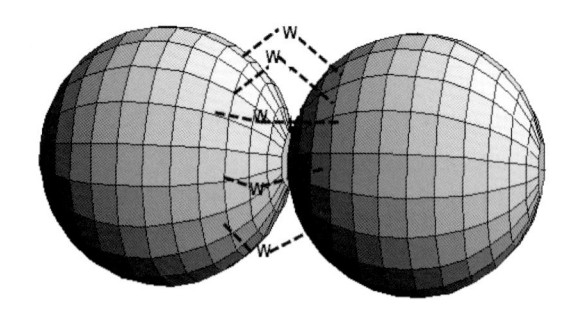

Fig. 3.11. Two proteins with water bridges connecting pairs of hydrophilic groups.

($-T\Delta S$) to the Gibbs energy. The energy change is large and negative, most likely due to hydrogen-bond bridges by water molecules. This is probably the main "thermodynamic force" for the dimerization process of proteins.

3.4.3 *A Simple Chemical Equilibrium*

One of the greatest successes of the free-energy formulation of the Second Law is the possibility of calculating the equilibrium constant for a chemical reaction. Such calculations are enormously useful in many industries as well as in chemistry, biochemistry, and medicine.

Here we present a very basic result of this kind of application of the Second Law.

Consider a system of N molecules in an ideal gas phase at some temperature T and pressure P. For simplicity, we assume that each molecule can be in only two energy levels, E_H and E_L, and the corresponding degeneracies are ω_H and ω_L, respectively (see Figure 3.12).

We can now view this system as a mixture of two species: H-molecules and L-molecules. The question we ask is, "What are the concentrations of each of the species H and L at equilibrium?"

Anyone who knows statistical thermodynamics would easily answer this question by invoking the Boltzmann's distribution, which states that the probability of finding a molecule in state H and L is given by $\Pr(H)$ and $\Pr(L)$ in Box 3.1 [For details see Ben-Naim (1992).]

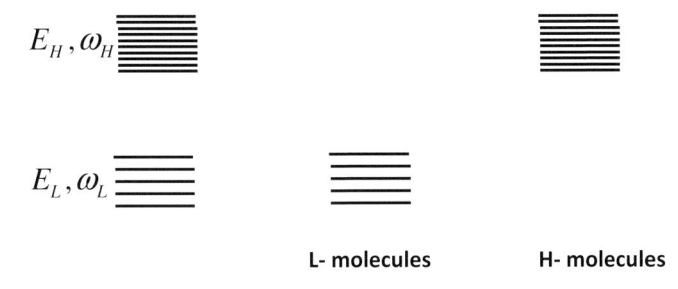

Fig. 3.12. A molecule with two energy levels, E_H and E_L, may be viewed as a mixture of two components.

182 | Entropy: The Truth, the Whole Truth and Nothing But the Truth

Box 3.1 The Boltzmann distribution for the two species H and L is

$$\Pr(H) = \frac{\omega_H \exp[-E_H/k_B T]}{C}$$

$$\Pr(L) = \frac{\omega_L \exp[-E_L/k_B T]}{C}$$

And

$$C = \omega_H \exp[-E_H/k_B T] + \omega_L \exp[-E_L/k_B T]$$

This is of course true. It seems as if this result has nothing to do with the Second Law. However, underlying the Boltzmann's distribution is the more fundamental principle that the Gibbs energy of the system (at P, T, and N constant) must be a minimum at equilibrium. This is essentially the Gibbs energy formulation of the Second Law. When applied to any chemical equilibrium, it provides an expression for the equilibrium constant for the reaction. For our particular example, the condition of minimum Gibbs energy leads to the equilibrium constant shown in Box 3.2.

Box 3.2 Minimizing the Gibbs energy G with respect to x_H, at constant T, P, N gives the fractions of H and L and the equilibrium constant

$$K = \frac{x_H^{eq}}{x_L^{eq}} = \exp[-\Delta G°/k_B T]$$

$$\Delta G° = \Delta E° - T\Delta S°$$

$$\Delta E° = E_H - E_L$$

$$\Delta S° = k_B \ln \frac{\omega_H}{\omega_L}$$

$$x_H^{eq} = \frac{K}{1+K} = \Pr(H)$$

$$x_L^{eq} = \frac{1}{1+K} = \Pr(L)$$

Applications and Misapplications of Entropy | 183

Note also that from the equilibrium constant K we can calculate the mole fractions of H and L at equilibrium, and these are equal to the Boltzmann's probabilities in Box 3.1. Thus, we conclude that the Boltzmann distribution is a result of the Second Law of Thermodynamics.

It is also of interest to note that in this particular example the standard Gibbs energy is determined by the "competition" between the energy difference $\Delta E°$ and the ratio of the degeneracies in $\Delta S°$. At very low temperatures ($T \to 0$) the entropy term $T\Delta S°$ is negligible, and the equilibrium constant is determined by the energy difference. In this limit all the molecules will be in the L state, i.e. $x_L^{eq} \to 1$, $x_H^{eq} \to 0$. On the other hand, for very high temperatures ($T \to \infty$), the entropy term dominates the standard Gibbs energy. In this limit, whatever the energy difference is, the mole fractions at equilibrium will be $x_L^{eq} = \frac{\omega_L}{\omega_L + \omega_H}$, $x_H^{eq} = \frac{\omega_H}{\omega_L + \omega_H}$.

We conclude by reiterating the role of the Second Law in this particular example. Suppose we start with any arbitrary initial concentrations of H and L, say x_L and x_H in the presence of an inhibitor. We remove the inhibitor, keeping T, P, N constant. A spontaneous process will ensue such that $x_L \to x_L^{eq}$ and $x_H \to x_H^{eq}$. The entropy change could be negative or positive, but the Gibbs energy change will be *negative*. At equilibrium the Gibbs energy will be at a minimum with respect to the variable x_L (or $x_H = 1 - x_L$).

3.4.4 *Protein Folding*

Here we discuss another case of a chemical reaction which is more complicated than the case discussed in Section 3.4.3. The process is known as protein folding. This process is also considered to be one of the "big unknowns" of science.[6] We describe here very briefly the problem and suggest a possible solution. As in the case of Section 3.4.3 we apply here the Gibbs energy formulation of the Second Law. Unlike the case discussed in Section 3.4.3, here it is sometimes more difficult to identity the "parameter" with respect to which Gibbs energy has a minimum.

Proteins are essentially a linear sequences of 20 different amino acids. Once the protein is synthesized, it folds into a precise 3-D structure. In most cases it is known that in order to properly function (say, as an enzyme), the protein must attain a very specific 3-D structure.[7] In the following, we discuss protein folding process *in vitro*, not *in vivo*.[8]

Suppose that we start with a dilute system of a protein in an aqueous solution. The protein has a huge number of conformations. Each conformation is determined by a set of internal rotational angles $\phi_1, \ldots, \phi_n$. For simplicity, we first group all the possible conformations into two groups. One group, containing a relatively small number of conformations is referred to as the folded form and denoted by F. All the other conformations, are referred to as the unfolded form, denoted by U. Thus, the folding-unfolding reaction is viewed as an isomerization reaction:

$$U \rightleftharpoons F$$

Suppose we start with a dilute solution of protein molecules all in the state U, at some temperature T, pressure P, and solvent composition N. We assume that there is an inhibitor which inhibits the conversion between the two isomers U and F.

We remove the inhibitor (or add a catalyst) and we find that a spontaneous process occurs, and a new equilibrium state is reached in which some molecules are in the state U, and some in the state F. Let us denote by x_F the mole fraction of the F form, and $x_U = 1 - x_F$ the mole fraction of the U form.

The Second Law formulated in terms of the Gibbs energy states that at the new equilibrium state the Gibbs energy function attains a minimum. More precisely, the function $G(T, P, N; x_F)$ has a *single* minimum with respect to x_F keeping T, P, N constant. This principle determines the mole fraction x_F^{eq} at equilibrium.

The process of a spontaneous folding; i.e. a molecule "flowing" from the U to the F forms, involves a *negative* change in the Gibbs energy. As we have done in Section 3.4.2, we write $G = E + PV - TS$, and ignore the PV term in the present discussion. Thus, the change in the Gibbs energy for conversion of one molecule (or one mole of molecules) from the U state to the F state has two contributions, ΔE and $T\Delta S$. What are the molecular factors that determine the values of ΔE and ΔS?

As for the entropy, we can argue that ΔS for the folding process will be negative. The main reason for this is that, in the process of folding, the protein goes from a macrostate containing many microstates to a macrostate containing a much smaller number of microstates. This is similar to the dimerization process in which an atom of hydrogen that has initially access to the entire volume V, is now confined to the relatively small volume v (see Figure 3.9).

Thus, we can rationalize that the main contributor to the change in the entropy for this process is the reduction in the number of microstates. The interpretation of this negative entropy change is similar to the one discussed in Section 3.4.3. The uncertainty, or the SMI associated with the U form, is much larger than with the F form. (This is a very qualitative argument. There are many other contributions due to rotations, vibrations, and solvent rearrangements that will contribute to ΔS). Thus, ΔS for the folding is expected to be large and negative. Therefore, $-T\Delta S$ is large and positive.

At this point, we face the same puzzle that we encountered in the case of dimerization of proteins. We know that the protein folds spontaneously (at some T, P, and N). Therefore, ΔG must be *negative*. We just found that one contribution to ΔG is large and *positive*. Therefore, the question is: What is the molecular origin of the net negative ΔG? More specifically, we must explain the source of the large negative *energy* change which is large enough to overcompensate for the positive change of $-T\Delta S$.

As in the case of protein-protein association, the molecular factors which are responsible for making the total Gibbs energy change negative are highly controversial. Most people believe that it is the *hydrophobic effect*. This is essentially the removal of hydrophobic groups from being exposed to water into the interior of the protein (see Figure 3.13). My view is radically different. The main factors

(a)

Transfer of a hydrophobic
molecule from water
into an organic liquid

(b)

Transfer of a hydrophobic
Group (blue) from water
into the interior of the protein

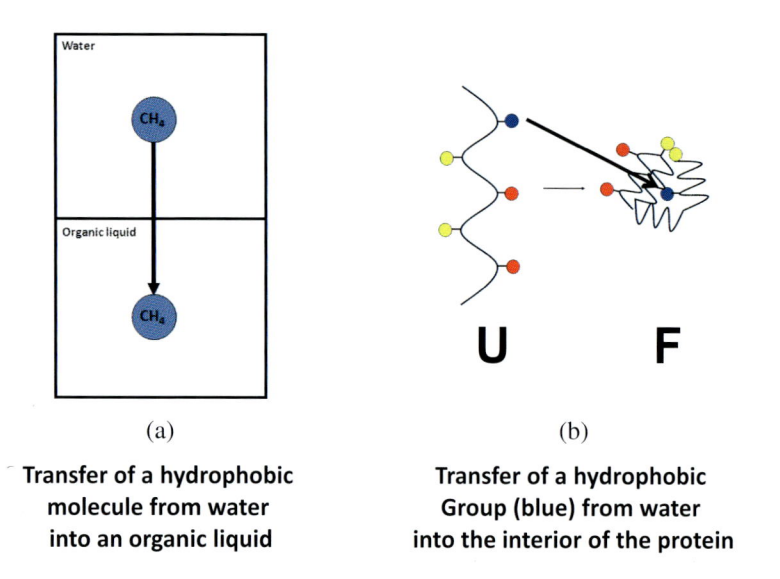

Fig. 3.13. Schematic description of the hydrophobic effect: (a) transfer of a hydrophobic molecule from water into an organic liquid, and (b) transfer of a hydrophobic group from water into the interior of the protein.

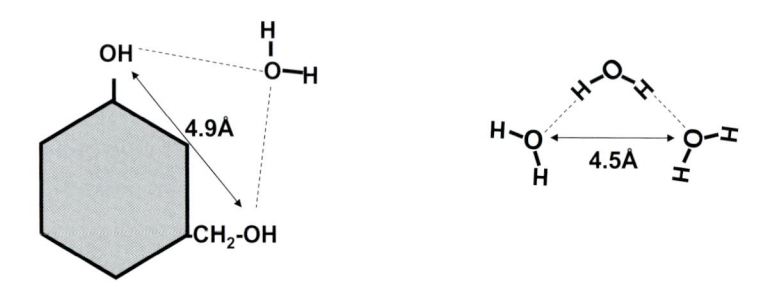

Fig. 3.14. Schematic description of the hydrophilic effect.

contributing to the negative Gibbs energy change are the *intramolecular* hydrogen bonding, as well as what I call *hydrophilic interactions*. These interactions are due to water molecules forming hydrogen-bond bridges between two or more hydrophilic groups which are exposed to the solvent. These are shown schematically in Figure 3.14. More details on this problem are given in Ben-Naim (2013, 2016b).

To summarize, proteins fold spontaneously in aqueous solutions. In this process, at T, P, N constant, the entropy change is negative. The thermodynamic potential is the Gibbs energy. The change in the Gibbs energy is negative in spite of the positive contribution of $-T\Delta S$.

So far, we have viewed the protein solution as a mixture of two isomers U and F at chemical equilibrium. One can also view each of the conformations of the protein, specified by the internal rotational angles $\phi_1, \ldots, \phi_n$, as a single species or a single isomer. In this view, we have an infinite number of isomers, a few of them considered to be of the F form, or the native structure of protein.

In this view, we find a very common misuse of the Second Law. Some people claim that the Second Law implies that the native structure of the protein "resides" in the global minimum of the Gibbs energy function $G(T, P, N; \phi_1, \phi_2, \ldots, \phi_n)$. Unfortunately, this formulation of the Second Law is incorrect. For more details, the reader is referred to Ben-Naim (2016b).

3.5 Entropy Change in Phase Transitions

This is a very interesting example showing how the entropy changes determine the boundaries of phases in a phase diagram.

Figure 3.15 shows a typical phase diagram of a substance. The three solid curves are the coexistence curves for the solid-liquid (SL), solid-gas (SG), and liquid-gas (LG).

For a single phase of a one-component system, the pressure and the temperature can be changed independently (within certain regions). This is shown by the two arrows ΔP and ΔT in the region when only pure liquid exists.

On the other hand, when there are two phases at equilibrium there is a unique functional dependence between the pressure and the temperature, i.e. given the temperature, the pressure is determined. These are the coexistence curves shown in Figure 3.15.

The slopes of all these curves are determined by the ratio of the change in entropy (ΔS) and the change in volume (ΔV), for the

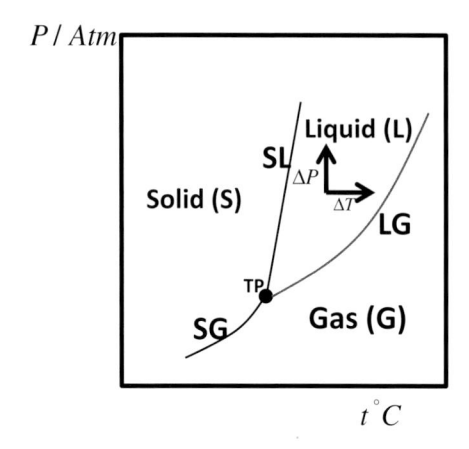

Fig. 3.15. Typical phase diagram showing the regions of solid liquid and gas. TP is the triple point.

corresponding change from one phase to another.

$$\frac{dP}{dT} = \frac{\Delta S}{\Delta V}$$

This equation is referred to as the Clapeyron equation.[9]

As one can see from Figure 3.15, the slopes of the three curves are all positive. For instance, the entropy change of melting and the volume change of melting are both positive. In this particular case, one can interpret the entropy of melting ΔS_m as due to transition from an "ordered" phase (solid) to a disordered phase (the liquid). However, a better interpretation is obtained by using the relationship $\Delta S_m = \Delta H_m/T$ where ΔH_m is the melting enthalpy. Thus, the positive change in entropy is interpreted in terms of a positive change in enthalpy — i.e. heat must be supplied to transform, say, one mole of solid into a liquid, at equilibrium.

The same is true for the transition between the solid and the gaseous phase; both ΔS_s and ΔV_s of sublimation are positive. Again, one may interpret ΔS_s as due to transition from an "ordered" phase (solid) to a disordered phase (gaseous). However, a better interpretation of the entropy change would be in terms of the heat of sublimation, i.e. $\Delta S_s = \Delta H_s/T$.

Finally, we discuss the liquid-gas transition which again is characterized by a positive slope. Here again, the volume change of evaporation is always positive, $\Delta V_v > 0$, and the corresponding change in entropy is also positive, $\Delta S_v > 0$. However, in this case we can hardly claim that the positive change in entropy is due to transition from an "ordered" (liquid) to a disordered phase (gas). More appropriately, we should interpret the positive entropy of vaporation in terms of a positive *heat* of evaporation; the transition from the liquid to the gaseous phase always involves positive heat transfer. The reason is that one needs to invest energy in order to achieve that transfer, from a phase in which the average interaction energy among the molecule is large, to a phase in which the average interaction energy is much smaller.

Thus, in all of the three transitions discussed above, ΔS as well as ΔV are positive. This in general is not true for transitions between different solid phases.

3.5.1 *Phase Diagram of Water*

Figure 3.16 shows the phase diagram of water at moderate pressures. It is seen that the solid-liquid coexisting curve has a negative slope. This is very unusual phenomenon. Most substances show a positive slope for the solid-liquid phase (see for example, Figure 3.15). From the Clapeyron equation we might suspect that perhaps the entropy change in the melting of ice is negative. This is, however, not the case. Ice has a molar volume *larger* than water. This is a well-known anomaly of water. Therefore, when ice melts the change of volume per mole is negative, $\Delta V < 0$. This is the reason for the negative slope of the solid-liquid curve.

Figure 3.17 shows the phase diagram of water at very high pressures. Note that the entire phase diagram shown in Figure 3.16 is not seen here. The reason is that in Figure 3.17 the units of pressure are in

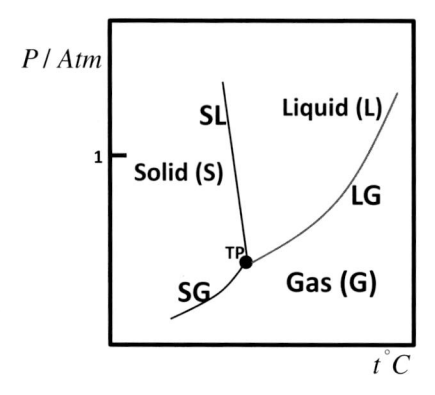

Fig. 3.16. Schematic phase diagram of water, showing the regions of solid, liquid and gas. TP is the triple point.

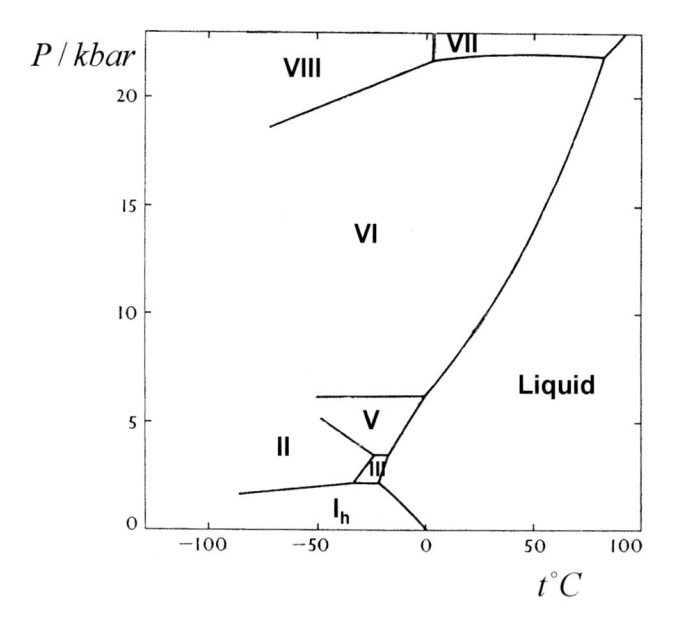

Fig. 3.17. Phase diagram of water at very high pressures.

kbar (about 1,000 atm). Thus, even one kbar is already much higher than the whole range of pressures shown in Figure 3.16. Note that in Figure 3.17 we see at least eight stable phases of ice. The one denoted by I_h is the ordinary, hexagonal ice.

There are many interesting phenomena associated with the transitions between the various solid phases of ice. For instance, the entropy changes for the transitions between I to III, III to V, V to VI, and VI to VII, are very small. This is the reason why the corresponding coexisting curves are almost horizontal (i.e. slope nearly zero). On the other hand, some of the entropy changes of transitions are of the order of 0.8 cal/mol K. This is roughly the same as the residual entropy of ordinary ice (see Section 3.2). It is assumed that in the ices I_h, III, V, VI, and VII the hydrogen atoms have the same configurational degeneracy we calculated in Section 3.2. Thus, the zero slope, say, between ice V and VI, is explained as a result of zero change in entropy. On the other hand, the transition from ice I to II has an entropy change of about -0.76, which is about -0.8, indicating that in ice II, the hydrogen atoms are *ordered*, and there is configurational entropy due to the arrangements of the hydrogen atoms as we have calculated for ice I. There are many other interesting properties of the different solid phases of ice. For more details, see Eisenberg and Kauzmann (1959).

3.6 Trouton's Law

Trouton's Law is an empirical law; it states that the entropy of vaporization, at one atmospheric pressure, of many liquids is almost constant, about $\Delta S_{vap} \approx 85\text{--}87/\text{J mol}^{-1}\text{ K}^{-1}$. Table 3.4 shows some values of the entropy of vaporization of several substances. The entropy of vaporization is the change of entropy for the transfer of one mole of the substance from the liquid phase to the gaseous phase when the two phases are at equilibrium.

In Section 3.4, we saw that the entropy of vaporization, together with the volume of vaporization, determines the slope of the $P(T)$ curve in the phase diagram. Here, we note that at equilibrium between the two phases, the Gibbs energy of vaporization is zero, $\Delta G_{vap} = \Delta H_{vap} - T\Delta S_{vap} = 0$. Therefore, $T\Delta S_{vap} = \Delta H_{vap}$, where ΔH_{vap} is the heat of vaporization. The last equality is used to determine ΔS_{vap},

Table 3.4 Entropies of Vaporization of Liquids at their Normal Boiling Point

	$\Delta S_{vap}/\text{J mol}^{-1}\text{K}^{-1}$
Benzene	+87.2
Carbon disulfide	+83.7
Carbon tetrachloride	+85.8
Cyclohexane	+85.1
Decane	+86.7
Dimethyl ether	+86.0
Methane	+73.2
Methanol	+104.1
Ethanol	+110.0
Water	+109.1

i.e. one measures the heat of vaporization dividing by the temperature to get ΔS_{vap}.

Table 3.4 shows values of the entropies of vaporization of some liquids at their normal boiling point (i.e. the boiling temperature at 1 atm pressure). The first six liquids in the table show values of ΔS_{vap} within 85–87 J mol^{-1} K^{-1}. However, there are exceptions. When the interaction energies among the molecules are relatively weak, as in the case of methane, the value of ΔS_{vap} is smaller. On the other hand, for liquids with stronger intermolecular interactions, the values of ΔS_{vap} are much larger.

The large value of ΔS_{vap} of water is usually ascribed to the large "loss of structure" of water upon vaporization. While it is true that water is considered to be a structured liquid, it is better to ascribe the large value of ΔS_{vap} to the strong intermolecular interactions (hydrogen bonds) between the water molecules.

Note that ΔS_{vap} of ethanol and methanol is nearly the same as for water, but one cannot claim that these liquids are as "structured" as water.

3.7 The Entropy of Solvation of Argon in Water

I will discuss in this section what I believe is one of the most interesting and exciting examples where entropy changes reveal one of the unusual, some even say, anomalous, properties of water.

In 1945 Frank and Evans published an article on the thermodynamics of solvation of inert solutes in water and in other liquids. They found that the entropy of solvation of these solutes is much larger and negative in water, as compared with the entropy of solvation of the same solutes in other liquids. In order to explain these findings, the authors conjectured that when an inert solute dissolves in water, it forms some kind of "iceberg" around it. This idea has captured the imagination of many scientists for more than half a century. How can an inert solute, weakly interacting with water molecules, form an "iceberg"? The contrast between the "innocent" small atom of argon or neon and the "magnificent" structured icebergs, must have impressed many scientists, including me. This was also the reason I chose to study the thermodynamics of solvation of inert gases in aqueous solutions in my PhD thesis.

Frank and Evans did not prove that an inert solute builds up an iceberg around it, nor did they provide any explanation as to why an inert solute should form icebergs. All they did was to interpret the large negative change in the solvation entropy in terms of increasing the order, or equivalently, increasing the structure of water. Yet, this idea was used by many scientists to explain the entropy and the enthalpy of solvation of the non-polar solute in water. In addition, this idea was overused in explaining other phenomena. We show here that the idea of iceberg formation, or a variation of this idea, can explain the outstanding large and negative entropy and enthalpy of solvation of inert solutes in water, provided one first explains how a simple solute forms icebergs. However, the idea of iceberg formation cannot explain the Gibbs energy of solvation of such solutes, nor the hydrophobic effect [see Ben-Naim (2011a, 2016b)].

Table 3.5 Values of the Solvation Gibbs Energy, Entropy, and Enthalpy of Methane in Water and in some Non-aqueous Solvents at Two Temperatures (Ben-Naim, 2009)

Solvent	$t,°C$	ΔG_s^* cal mol^{-1}	ΔS_s^* cal mol^{-1}K^{-1}	ΔH_s^* cal mol^{-1}
Water	10	1,747	−18.3	−3,430
	25	2,000	−15.5	−2,610
Methanol	10	343	−2.6	−390
	25	390	−3.7	−710
Ethanol	10	330	−3.2	−570
	25	380	−3.5	−650
1-Propanol	10	345	−4.3	−880
	25	400	−3.0	−500
1-Butanol	10	369	−2.8	−420
	25	430	−4.5	−910
1-Pentanol	10	399	−3.3	−530
	25	450	−3.6	−630
1,4-Dioxane	10	538	−0.8	+310
	25	553	−1.1	+220
Cyclohexane	10	154	−1.9	−390
	25	179	−1.4	−230

Table 3.5 shows the solvation Gibbs energy, entropy, and enthalpy of methane in water and in some organic liquids. Figure 3.18 shows the entropy and enthalpy of solvation of xenon in water and some linear alcohols.

As can be seen from Table 3.5, the Gibbs energies of solvation of the non-polar solutes in water are positive and much larger than in other liquids. On the other hand, the entropies and enthalpies of solvation in water are large and negative in water as compared with the non-aqueous solvents.

By the process of *solvation*, we mean here the transfer of a molecule from a fixed position in an ideal gas phase to a fixed position in the liquid (see Figure 3.19). The values of ΔS_s^* in Table 3.5 reveal

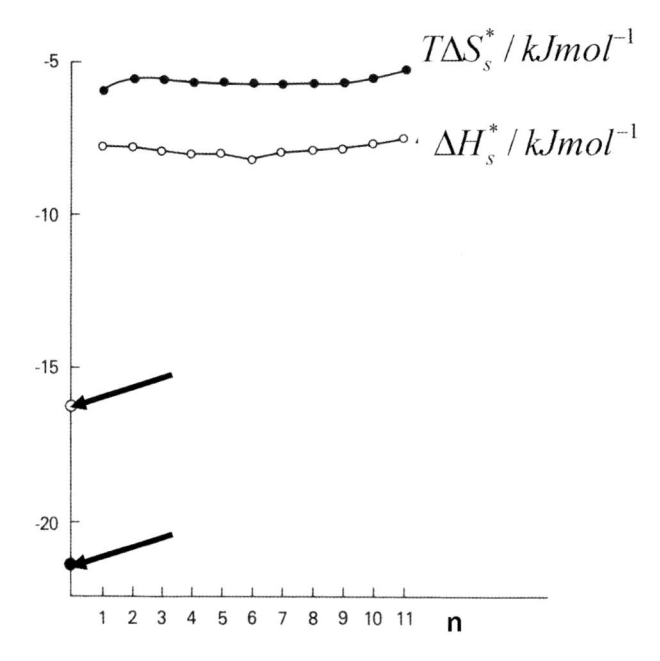

Fig. 3.18. Solvation enthalpy $\Delta H_s^*/\text{kJ mol}^{-1}$ (open circles) and entropy $T\Delta S_s^*/\text{kJ mol}^{-1}$ (full circles) for xenon in a series of linear alcohols; n is the number of carbon atoms in the alcohol. The experimental values for water are shown as open and full circles at $n = 0$. All values are for 1 Atmospheric pressure and 20°C.

that the solvation entropy of methane in water is much larger (in absolute value) than in any other organic liquid. In order to understand these values, let us consider first the solvation of KCl in water.

Figure 3.20 shows schematically the arrangement of water molecules around two ions, K$^+$ and Cl$^-$. It is known that the solvation entropy of KCl in water is large and negative. Now, consider the following "reaction":

$$K^+ + Cl^- \rightarrow K + Cl \rightarrow A + A$$

In this process we first transfer the electron from the chlorine ion to the potassium ion. This will produce two neutralized atoms K and Cl. Next, we replace the two atoms K and Cl by two argon atoms A.

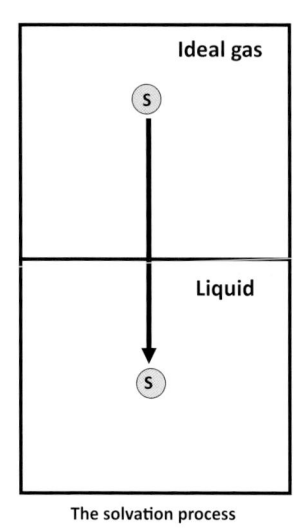

Fig. 3.19. Definition of the solvation process.

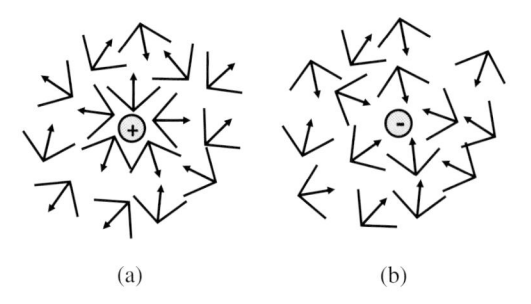

Fig. 3.20. Arrangement of water molecules around a (a) positive, and (b) negative ion. The arrows indicate the dipole moment of the water molecule.

It is relatively easy to understand why the entropy of solvation of KCl in water is large and negative. The electric field around the charged ion is very strong. This field will force the dipole moments of the water molecule to orient themselves as shown in Figure 3.20. Accepting for the moment the interpretation of entropy as a measure of order, we can associate the negative solvation entropy of KCl with this preferential orientation, or *ordering*, of water molecules around the ions. More appropriately, we can say that the range of accessibility of both locations and orientations of the water molecules is reduced.

Once we remove the charges on the ions, we also "turn off" the electric field around the ions. Therefore, we cannot expect any "ordering" effect around the ions, hence we expect that the entropy of solvation of the two neutral atoms will be smaller (in absolute values) than for KCl. Equivalently, we should expect that in the "reaction" of neutralization of the ions, the entropy change will be positive. Figure 3.21 shows a schematic random arrangement of water molecules around the argon atom.

The (surprising) experimental fact is that the entropy of solvation of two argon atoms (roughly having the same size as the neutral atoms K and Cl) is even more negative than the entropy of solvation of KCl. The same is true for two methane molecules.

These findings are quite puzzling. Why should the entropy of solvation of two neutral atoms be more negative than that of the charged atoms?

The large negative entropy of solvation of a simple inert solute was quite striking and difficult to explain. This fact was even more surprising when it was compared with the entropy of solvation of two ions such as K^+ and Cl^-, which have roughly the same size as argon atoms. The main problem is how to explain these outstanding values

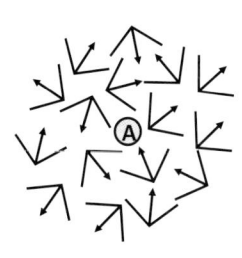

Fig. 3.21. A random arrangement of water molecules around an argon atom.

of the entropy and the enthalpy of solvation of the inert solutes in water on a molecular level.

Two theories were suggested. The first was by Eley (1939, 1944), which was based on a lattice theory of water (a very common approach to liquids at that time). Eley assumed that there are a fixed number of holes, or cavities, in liquid water. When an inert gas dissolves in water, it is confined to these fixed numbers of holes. In his view, the reduction in the translational entropy of the molecules in the process of solvation explains the negative entropy change. A completely different picture was suggested by Frank and Evans (1945). These authors did not assume any fixed number of holes, nor a fixed structure of the water. Instead, they proposed that the dissolved molecules create some kind of a new structure in water, referred to as "icebergs."

It should be noted that at the time of the publication of Frank and Evan's article the concept of entropy was understood as a measure of order or disorder. The negative entropy of solvation of non-polar molecules was interpreted as an increase in order, or equivalently as an increase in the structure of water. Frank and Evans did not explain why an inert solute would cause an increase in the structure. They simply translated the experimental fact of a negative change in entropy, into the language of increasing the order, or the structure, of the system.

Both theories tried to explain the negative entropy change in terms of increase in order. Confining solute molecules to a fixed number of holes was clearly conceived as a more "ordered" state, compared with that of molecules in the gaseous phase. The smaller the number of holes, the smaller the volume accessible to the molecules. Hence the entropy of solvation is expected to be negative. On the other hand, Frank and Evans' picture relegates the "ordering" effect to the water, i.e. building up *new* structures.

Eley's theory did not survive long. The idea of water having a fixed number of holes was untenable. On the other hand, Frank and Evans' idea has survived until today. It survived not because it was proven to be correct, but rather because it was a nice pictorial idea that captured the imagination of most scientists working in the field of aqueous solutions.

Not only did the iceberg idea survive, but it was used over the years to explain many other phenomena associated with aqueous solutions. It was much later that the structuring of water by an inert solute was reformulated, not in terms of building up a new structure (as in the case of ionic solutes, or what Frank and Evans envisaged by icebergs), but in terms of *enhancement* of the already existing structure of liquid water. This was only a shift in formulating the problem. It was not yet an answer to the question why an inert solute would enhance the structure of water. It was only much later that a firm argument based on the Kirkwood-Buff theory was provided [see Ben-Naim (1992, 2006a)].

Thus, the main idea that argon or methane can enhance the structure of water was found to be correct. In order for this to occur, one must assume that pure liquid water must have a large degree of structure before the introduction of the argon or the methane molecules. The fact that an enhancement of this structure is possible is still considered one of the most outstanding properties of liquid water. For more details, see Ben-Naim (1992, 2011a).

3.8 "Entropy of Mixing"

Mixing of two different ideal gases, as illustrated in Figure 2.9a, is perhaps the simplest example for which the entropy change can easily be calculated. It is also a process which is misunderstood by most authors who write about it, including Gibbs, who was the first to thoroughly study this process (see below). If we have one mole of A in a volume V, at a temperature T, and another mole B in a volume V, and temperature T, then the entropy change for the process shown in Figure 2.9a is

$$\Delta S = 2R \ln 2$$

While there is no question about the value of ΔS for this process, there is great confusion regarding its interpretation.[10] Here are three main interpretations of this increase in the entropy.[11]

(i) The experimental interpretation

Mixing is viewed as a process of disordering. We all agree that the mixture of A and B on the right-hand side of Figure 2.9a is a more disordered state than the system on the left-hand side of Figure 2.9a, where the two components are separated.

Next, most textbooks interpret entropy as a measure of disorder. Positive change in entropy means that the system became more disordered. Therefore, one concludes that the entropy change in the mixing process is due to the mixing, i.e. to increasing disorder, hence, the term "entropy of mixing." This interpretation sounds very reasonable. Nevertheless, it is wrong!

(ii) The statistical mechanical interpretation

Using the methods of statistical mechanics of ideal gases, it is easy to calculate the entropy change for the process as shown in Figure 2.9a (for details, see Ben-Naim, 2008). The result is

$$\Delta S = 2R \ln \frac{2V}{V} = 2R \ln 2.$$

This is exactly the same result we obtain from thermodynamics. However, in this calculation the number 2 under the logarithm is a result of the increase in the accessible volume for each of the $2N$ particles in the system, not a result of the mixing.

(iii) The informational interpretation

This is essentially the same as the statistical mechanical interpretation. However, now we calculate the difference in the SMI for the two states of the system. If we use the logarithm to the base 2 we get

$$\text{change of SMI} = 2N \log_2 2 = 2N \text{ bits.}$$

We have seen in Chapter 1 that the entropy is the same as the SMI for equilibrium states (after multiplying by $k_B \ln 2$). We find that the change in the SMI in this process is $2N$ bits, i.e. one bit per particle. Each particle was initially located in volume V, and finally in volume $2V$. Therefore, there is loss of information of one bit per particle. This change in the SMI, as well as in entropy, is independent of the type of the particles. In our case it is one bit for each A molecule, and one bit for each B molecule.

The last conclusion is not trivial. Gibbs, who was the first to analyze the so-called "entropy of mixing," was puzzled by the fact that the entropy of mixing is independent of the kind of molecules, for as long as A and B are distinguishable molecules. Note that this conclusion is valid for ideal gases only. If you mix two liquids, two solids, or two non-ideal gases, you will find that the entropy change in the mixing process *does* depend on the kind of molecules A and B. In the latter cases the entropy of mixing is a result of two factors: one is the change in the volume accessible to each particle, while the second is due to the change in the interaction energies for the pairs A-A, A-B, and B-B.

When we mix two ideal gases, we completely neglect all intermolecular interactions. In this case the entropy change in the mixing process shown in Figure 2.9a is due only to the change in the accessible volume. Hence, the so-called "entropy of mixing" of ideal gases is nothing but the entropy of expansion; each gas expands from V to $2V$. In this view there is nothing puzzling in the fact that the "entropy of mixing" of ideal gases is independent of the kind of molecules A and B. It is unfortunate that Gibbs himself failed to see that mixing of two different ideal gases is nothing but expansion of each gas from V to $2V$, and that the mixing, by itself, does not contribute anything to the thermodynamics of mixing (of ideal gases). Mixing, as well as demixing, of ideal gases, can occur with positive, negative, or zero change in entropy.

There is another process for which Gibbs reached a wrong conclusion. This is the process shown in Figure 2.9b. This process is exactly the same as in 2.9a, except that in this case the two compartments contain the same molecules, say A. In this process, there is no change in entropy when we remove the partition between the two compartments. This result ($\Delta S = 0$) is accepted by everyone. There is, however, disagreement regarding the interpretation of this result. Here are the three interpretations of the process shown in Figure 2.9b.

(i) The thermodynamic or experimental interpretation

Here, the result $\Delta S = 0$ is straightforward, if not obvious. After removing the partition in Figure 2.9b, we do not observe any process. If we do not see anything happening, then it is natural to conclude that there is no process. If there is no process, then there should be no change in entropy: an obvious conclusion. Obvious indeed, but a wrong conclusion!

(ii) The statistical mechanical interpretation

If we write the partition function for the initial and the final states, we can calculate the change in the Helmholtz energy for this process, and from its temperature derivative, we can calculate the entropy change [for details, see Ben-Naim (2008)]. The result is

$$\Delta S = 2Nk_B \ln 2 + k_B \ln \frac{(N!)^2}{(2N)!}.$$

One can easily prove that this entropy change must always be positive.[12]

Thus, statistical mechanics tells us that there are two processes going on in the seemingly non-process shown in Figure 2.9b. One is expansion from V to $2V$, the second is the change in the number of indistinguishable particles. We started with N indistinguishable particles in the left compartment, and another N indistinguishable particles in the right compartment. After removing the partition between the two compartments, we have $2N$ indistinguishable particles. Thus, we have two contributions to the entropy changes; one due to the expansion of the gases, and the second due to the change in the number of indistinguishable particles.

But, why is $\Delta S > 0$, in disagreement with the experimental result that $\Delta S = 0$? The answer is that $\Delta S > 0$ for any finite number of particles. However, when N is very large, such that we can use the Stirling approximation for $\ln N!$, the two terms cancel each other.[13]

Thus, we see that the two contributions to $\Delta S > 0$ for this process cancels out for large N, resulting in $\Delta S \approx 0$.

(iii) **The informational interpretation**

This interpretation is essentially the same as the statistical interpretation. However, conceptually it is slightly different. As for the mixing process shown in Figure 2.9a, we lose one bit per particle due to the expansion process. On the other hand, we gain one bit per particle due to the change in the number of indistinguishable particles. The net change is zero bits per particle (provided that N is very large).

Gibbs' understanding of the difference in the two processes in Figure 2.9

As we noted above, Gibbs was puzzled by the fact that the entropy of mixing of two different gases is the same, independent of the kind of molecules in the two compartments. As long as the two gases are different, there is a finite change in entropy. When the two gases are the same, there is no change in entropy. It should be said that Gibbs understood that the difference in the entropy changes for the two processes in Figure 2.9 is a result of the indistinguishability of the molecules. He did not see any paradox in the fact that ΔS changes discontinuously when we do the experiment with different kinds of particles, or with the same kind of particles. Later, people were puzzled by this discontinuous "jump" in ΔS from the finite value of $2R \ln 2$ to zero. This puzzle is sometimes referred to as the *Gibbs paradox*. Unfortunately, there has never been a paradox, and even Gibbs himself did not see it as a paradox. Gibbs understood that atoms and molecules can be either distinguishable or indistinguishable. There is no continuous change from being distinguishable to being indistinguishable. Such a continuous change could be envisaged for two *different* objects, say two labeled balls. One can imagine that the label is being continuously reduced to zero. This process will transform two different macroscopic objects into two identical objects. However, such a transformation cannot be achieved in the microscopic world of atoms and molecules. These particles are either distinguishable or indistinguishable. There is no continuous passage from one to another.

Notwithstanding this fundamental aspect of the microscopic world, Gibbs erred in his conclusion regarding the two processes in Figure 2.9. For the proper mixing in Figure 2.9a, Gibbs wrote that although it is an irreversible process, it can be reversed. Of course, one needs to invest energy to do so. The process in Figure 2.9a will not be reversed spontaneously.

On the other hand, for the process in Figure 2.9b, Gibbs wrote that its reversal is "entirely impossible."[14] The reason is clear. Once we remove the partition between the two compartments, each particle can wander in the entire volume $2V$. There is no way to "reverse" the process, in the sense of bringing back each particle originating from the left compartment to the same compartment; the same goes for the particles originating from the right compartment. Why? Because the particles are *indistinguishable*. Once we remove the partition there is no way we can tell which particle originated from which compartment. Hence, Gibbs concluded that this reversal is "entirely impossible."

Ironically, Gibbs failed to see that for the same reason he claimed that the reversal of the process in Figure 2.9b is "entirely impossible," it is in fact, trivially possible. One can, effortlessly and with no investment of energy, reverse this process by simply putting the partition at its original place. By doing so, we will get two compartments having (nearly) the same number of particles. No one can claim that the final state is not the same as the original state we started in Figure 2.9b. The reason is simply the indistinguishability of the particles.

Mixing and assimilation of ideal gases

We have seen that the process of proper mixing in Figure 2.9a has two components: expansion and mixing, one contributing $2Nk_B \ln 2$ to the entropy change, and the second contributing nothing to the entropy change. In the second process to which Gibbs referred to as "mixing of gases of the same kind," there are also two components: one due to expansion which is $2Nk_B \ln 2$, and the second due to "mixing of the same kind," which for large N is approximately $-2Nk_B \ln 2$. We shall refer to the process of "mixing of the same kind" as an *assimilation* process.

We saw that the assimilation process involved a *negative* change of entropy. We can define a "pure" assimilation process as one in which neither the accessible volume nor the velocity distribution changes. Such a process of pure assimilation is shown in Figure 2.12a (compared to the pure mixing process shown in Figure 2.12b).

Suppose we consider the reverse of the assimilation process depicted in Figure 3.22b; we can calculate the entropy change and find $\Delta S = 2Nk_B \ln 2$, which is positive.

Remember that the Second Law states that in a spontaneous process occurring in an isolated system, the change in entropy must

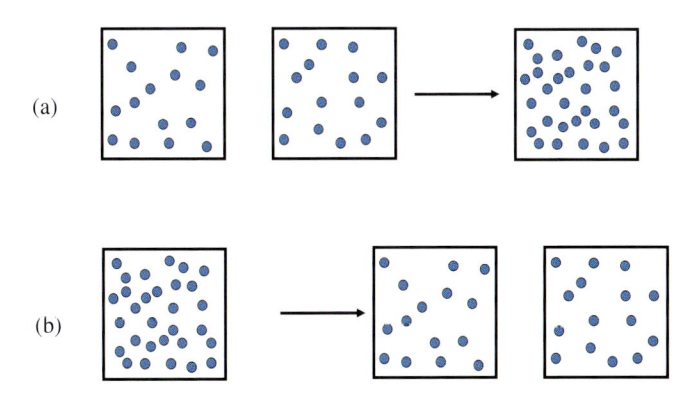

Fig. 3.22. (a) Pure assimilation process, and its reversal, and (b) pure deassimilation process.

be positive. A positive change in entropy, however, is not a sufficient condition for a spontaneous process to occur.

Here we have in Figure 3.22b a process for which we know that the entropy change is positive. However, this is obviously not a spontaneous process. We never observe a system of $2N$ particles in a box of volume V splitting into two boxes, each containing N particles in volume V.

The question posed now is the following: "Can we devise an experiment where a *spontaneous* pure *deassimilation* process occurs (i.e. equivalent to the process in Figure 3.22b) for which the entropy change is exactly $2Nk_B \ln 2$?

I believe the answer to this question is most exciting. We shall discuss such a process in the next section.

3.9 Racemization as a Pure Deassimilation Process

This section is devoted to the last example of a "legitimate" usage of entropy and the Second Law. It is the gem of all the examples. It is an example which cannot be explained by any of the common interpretations of entropy. It is the only pure deassimilation process which occurs spontaneously with positive change in entropy.

In Section 3.8 we described a process of pure deassimilation (see Figure 3.22b) for which $\Delta S = 2Nk_B \ln 2$. We asked whether there exists a spontaneous process of deassimilation occurring in an isolated system for which the entropy change is exactly $2Nk_B \ln 2$. By pure deassimilation process, we mean a process for which the accessible volume for each particle does not change, the velocity distribution (or the temperature) does not change, and the energy of the system does not change.

The question posed at the end of the previous section is, yes! The process is called *racemization*.

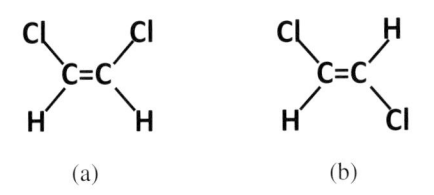

Applications and Misapplications of Entropy | 207

$$\underset{H}{\overset{Cl}{>}}C=C\underset{H}{\overset{Cl}{<}} \qquad \underset{H}{\overset{Cl}{>}}C=C\underset{Cl}{\overset{H}{<}}$$

(a) (b)

Fig. 3.23. Cis and trans isomers of dichloroethylene.

Consider first a chemical reaction of isomerization, i.e. a molecule having the same chemical formula, say, dichloroethylene (see Figure 3.23).

We start with one isomer, say the cis one on the left-hand side of Figure 3.23, and add a catalyst. The system will approach a new equilibrium state in which we shall have an equilibrium mixture of cis and trans. The entropy change in this process has two contributions: one, due to changes in the internal energies of the molecules, and the second, due to the change in the numbers of the cis and trans isomers.

There is one very special case of an isomerization reaction for which no changes in the internal energies of the molecules occur. This is the case of the two isomers which are mirror images of each other. These are referred to as two enantiomers. Figure 2.10 shows an example of such a molecule. A carbon atom has four different neighboring groups. Such a molecule is said to have a chiral center. As we can see from the Figure 2.10, the two isomers are identical except for being mirror images of each other.

The history of such molecules is truly facscinating.[15] Before we discuss the thermodynamics of such processes, I cannot resist a digression and tell you what I believe is the most interesting phenomenon in chemistry and biochemistry. During the early 19th century it was known that certain crystals rotate the plane of polarized light. However, it was Louis Pasteur who realized that the optical activity of compounds is associated with molecules having an asymmetric grouping of atoms. Here, we discuss only one class of such molecules which contain a chiral center, i.e. a carbon atom having four different groups attached to it, occupying the vertices of a regular tetrahedron.

The two molecules are mirror images of each other, and cannot be superimposed on each other as the right and the left hand (see Figure 2.10). This is why the chirality is associated with *handedness* (chiral means "hand" in Greek). Each isomer is referred to as an enantiomer, and the mixture of the two as a racemic mixture.

Many molecules in living systems are chiral. What is most interesting is that only one of the enantiomers is present; e.g. amino acids, the building blocks of proteins, are invariably the *l*-enantiomer, sugars are *d*-enantiomers, etc. Although these facts were known for over a hundred years, we still do not understand how nature "picked up" only one of the isomers. There are many conjectures regarding the origin and the sustenance of this puzzling phenomenon.

These two enantiomers have exactly the same structure, chemical formula, and properties, except for rotating the plane of polarized light in different directions. The two isomers are denoted *levo* (*l*) and *dextro* (*d*), for rotating to the left and right, respectively. Louis Pasteur (1840) was the first to resolve a mixture of two isomers (sodium ammonium salt of tartaric acid) by crystallization of a racemic mixture into two different crystals of the two (asymmetric) isomers (see Figure 3.24).

Actually, tartaric acid has two asymmetric or chiral centers as shown in Figure 3.24 (the "+" and "−" refer to the two optically active compounds, which as can be seen are mirror images of each other. The third structure also has two asymmetric carbon atoms, but the molecule as a whole is not asymmetric. It has a plane of symmetry bisecting the line connecting the two asymmetric carbon atoms. This isomer is called *meso*, and it is optically inactive).

When an enantiomer has greater affinity for molecules of the same kind, each of the two isomers will crystallize in its pure form. There are cases when there is greater affinity between the two different enantiomers. In this case the pairs of the two isomers, called *racemic compound* or *racemate*, form a crystal of 1:1 mixture of the

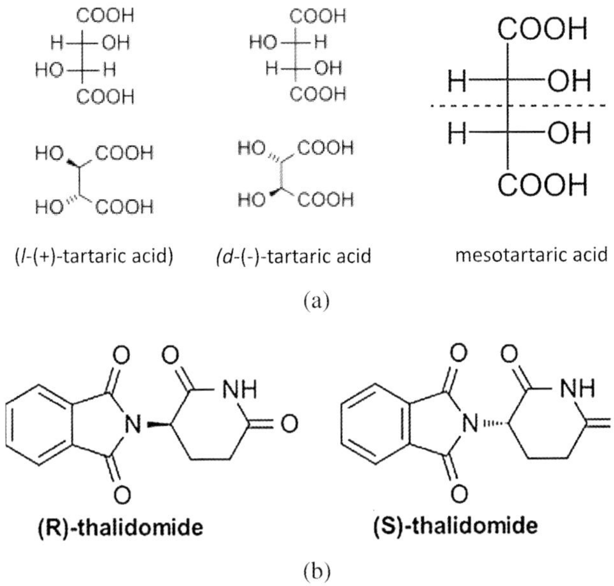

Applications and Misapplications of Entropy | 209

[Figure: (a) Tartaric acid stereoisomers — (l-(+)-tartaric acid), (d-(-)-tartaric acid), mesotartaric acid; (b) (R)-thalidomide and (S)-thalidomide]

Fig. 3.24. (a) Tartaric acid, and (b) thalidomide.

two isomers. This is the case for sodium ammonium salt of tartaric acid studied by Pasteur.

Over a hundred years after Pasteur's classical experiments, the drug company, Chemie Grünental, developed and sold a sedative drug based on the molecule called thalidomide (which was sold under different brand names). The drug sold over-the-counter was marketed as an anti-nausea, and to alleviate hyperemesis gravida rum, or what is commonly known as "morning sickness," in pregnant women.

Then came the so-called thalidomide tragedy when thousands of infants whose mothers took the drug were born with limb abnormalities. The precise mechanism of the action of thalidomide is not known. However, it was found out that only one of the enantiomers is the harmful component.

Let us go back to the entropy of assimilation-deassimilation process.

Suppose we start with pure l molecules. We add a catalyst which enables the conversion between the two isomers. At equilibrium we will find equal numbers of l and d molecules. Since the two isomers have the same internal energies, the entropy change for this reaction is due to only changes in the *number* of d and l molecules (here, from initially $2N, l$ molecules to finally, N, l and N, d molecules). The entropy change for this reaction is $2Nk_B \ln 2$. Here are three possible interpretations of this entropy change.

(i) **The thermodynamic interpretation**

The thermodynamic interpretation is that this entropy change is the same as the "entropy of mixing," i.e. the entropy change is due to the mixing of the d and the l isomers. Indeed, the quantity $2Nk_B \ln 2$ is exactly the same as the entropy change in the process shown in Figure 2.9a. As we have explained in Section 3.8, the term "entropy of mixing" is not appropriate for the process in Figure 2.9a. Although we observe a mixing of two components, the entropy change in the process of Figure 2.9a is *not* due to the mixing but to the expansion of the two gases from V to $2V$.

The term "entropy of mixing" is *a fortiori* inappropriate for the process of racemization. Here, we do not observe a mixing of two components, l and d, but an evolution of a pure l into a mixture of two components. Thus, thermodynamics cannot explain the origin of the racemization process.

Thermodynamics can only offer a qualitative explanation for the equal amounts of d and l at equilibrium. The equilibrium constant in this case is related to the standard Helmholtz energy for this reaction. Since the two isomers have the same internal energy levels, the standard Helmholtz energy of the conversion reaction ΔA° must be zero. Hence, the equilibrium constant must be one, hence N_l and N_d must be equal at equilibrium. The relevant equation is

$$K = \left(\frac{N_l}{N_d}\right)_{eq} = \exp\left[-\frac{\Delta A^\circ}{k_B T}\right]$$

(ii) The statistical mechanical interpretation

The statistical mechanical interpretation is simple. For an ideal gas, we can calculate the partition function in the initial and the final states, then calculate the entropy change. We find that the entropy change $2Nk_B \ln 2$ is due only to changes in the numbers of l and d molecules. As for the equilibrium ratio of N_l and N_d, statistical mechanics provides us with the equation

$$K = \left(\frac{N_l}{N_d}\right)_{eq} = \frac{q_l}{q_d} = 1$$

where q_l and q_d are the internal partition functions of l and d isomers, respectively. Since the internal energies of the two isomers are the same for the two isomers, we must have $q_l = q_d$ and therefore $K = 1$.

An interesting phenomenon is that the melting temperature of a racemic mixture is lower than the melting of the pure enantiomer. Figure 3.25a shows the phase diagram of such a case. When one cools a racemic mixture of d and l such that the affinity between dd and ll is larger than that of dl, we get a resolution, i.e. separation, into pure crystals of d and l. However, there are cases when the dl affinity is larger than dd (or ll). In such cases we get a solid consisting of pairs

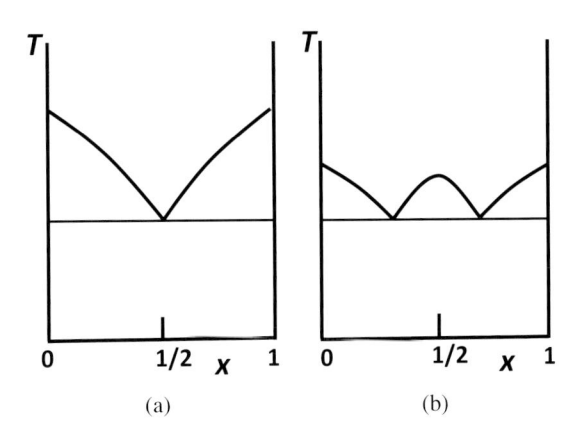

Fig. 3.25. Schematic phase diagram of two enantiomers: (a) one eutectic melting point, and (b) Two eutectic points.

of *d* and *l*, called racemate with composition 1:1. The phase diagram of such a case is shown in Figure 3.25b.

(iii) The informational interpretation

The informational interpretation is the same as the statistical interpretation, i.e. the change in entropy is due to changes in the identity of N molecules from *l* to *d*. We refer to this process as pure deassimilation process for the following reasons. First, the accessible volume for each particle does not change in the process (it is V initially and finally). Second, there are no energetic changes; neither the temperature nor the velocity distribution changes in the process. The only change that occurred is that half of the *l* molecules acquired a new identity *d*. To see that this is a deassimilation process as we have defined in Figure 3.22b, look at Figure 3.26a. We start with $2N$ molecules of *l*. We add a catalyst to initiate the conversion between *d* and *l*. After attaining equilibrium, we can remove the catalyst and demix the mixture of *l* and *d* (see Figure 3.26b). We get two boxes of equal volume, one having N molecules of the *l* form, and the second having N molecules

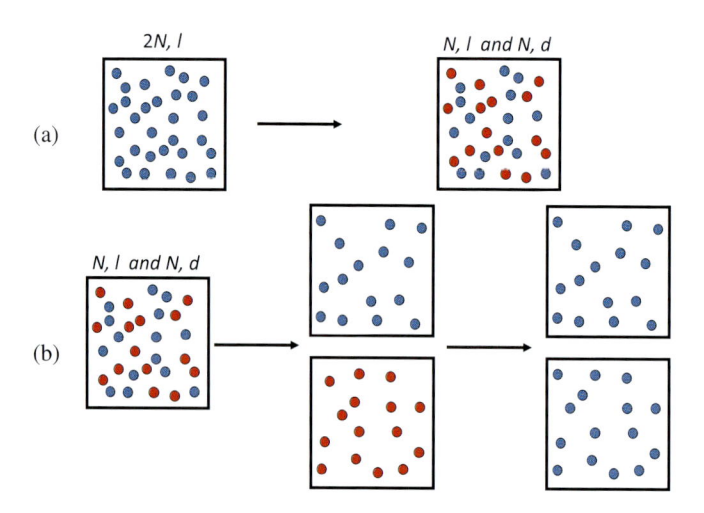

Fig. 3.26. (a) Racemization, and (b) equivalence of a racemic mixture and two systems of pure isomer.

Applications and Misapplications of Entropy | 213

of the d form. We already know that this process of demixing involves no change in entropy. Next, we convert (or if you like, replace) the box with N, d molecules into N, l molecules. Again, since the two isomers have the same internal energies, there is no change in either the energy or the entropy in this process. Thus, we end up with two boxes of equal volume, each having N molecule of the l isomer.

Now look at the entire process from the left-hand side of Figure 3.26a to the right-hand side of Figure 3.26b. You can see that this is exactly the reversal of the pure assimilation process shown in Figure 3.22b. Therefore, it is appropriate to refer to this process as *pure deassimilation*, and to ascribe the entropy change $2N_l k_B \ln 2$ to the deassimilation process.

To end this section, we also add an informational, theoretical interpretation for the eventual equilibrium state of l and d with the same concentration. The explanation is exactly the same as the one given in connection with the equilibrium state in the expansion process from V to $2V$.

Suppose that we prepare a mixture of l and d forms with an arbitrary composition. Let x_l and x_d be the mole fractions of l and d, respectively. Since we have $x_l + x_d = 1$, the pair (x_l, x_d) is a probability distribution.[16]

Therefore, on this probability distribution we can define the Shannon measure of information, $H(x_l, x_d)$. Now, we add a catalyst so that the conversion between l and d becomes possible. We can ask: What is the probability $\Pr(x_l, x_d)$ of observing any specific distribution (x_l, x_d)? It turns out that this probability distribution is related to the SMI defined on (x_l, x_d).[17]

Because of this relationship, the distribution which maximizes the SMI is the same distribution that maximizes the probability Pr (to which we referred as the *super probability* in Section 1.5.3). We expect that upon adding the catalyst, the system will evolve into the distribution which has maximum probability Pr. This is the same as the system with maximum SMI. As we know, the maximum SMI for a two-outcome experiment is obtained when $x_l = x_d = \frac{1}{2}$. Hence, we have an explanation for the final equilibrium concentration of *l* and *d*.

Thus we see that the mixture of two enantiomers is an excellent example of a process in which no volume, no temperature, and no internal energies are involved. It is a pure process of deassimilation. It is also an excellent example of demonstrating the workings of the Second Law of Thermodynamics.

I would like to conclude this section with a personal anecdote.

On numerous occasions when I gave lectures on the "entropy of mixing," I said that the entropy change for such a mixing process (see Figure 2.9a) is due to expansion, rather than to the mixing. I also introduced the example of the racemization to show that there exists a process of pure deassimilation which unfortunately, is interpreted in terms of "entropy of mixing."

In one of my lectures, someone in the audience stood up and said, "I have been teaching thermodynamics for so many years. I always taught that the 'entropy of mixing,' as well as the 'racemic process' are due to *mixing*, and I will continue to teach this process exactly the same way, simply because one cannot explain in a course of thermodynamics, the idea of assimilation or deassimilation."

My answer to this issue is simple. Of course, one cannot explain the process of assimilation within thermodynamics. One must use statistical mechanics for this purpose. However, the same is true for the entropy change of any process. One can never explain any entropy change by purely thermodynamic arguments.

3.10 Misusing Entropy in Explaining the Low Solubility of Argon in Water

In this section we begin with a few misuses of entropy and the Second Law. In Section 3.7 we discussed the interpretation of the large negative solvation energy of inert solutes in water. However, there is another outstanding phenomenon related to aqueous solutions of inert solutes. Table 3.5 shows that the Gibbs energy of solvation of methane in water is much larger than the solvation Gibbs energy of methane in other liquids. The Gibbs energy of solvation is related to the solubility of the gas; the larger the value of ΔG_s^* the smaller the solubility.[18] This phenomenon is referred to as one of the *hydrophobic effects*.

In many articles, as well as in many biochemistry textbooks, one finds statements alluding to an "explanation" of the hydrophobic effect by invoking the idea of the iceberg formation. Here, by "hydrophobic effect," we mean the *large positive* Gibbs energy of solvation of say, methane in water (see Table 3.5). We present here only briefly the main argument.

(i) The solvation entropy dominates the solvation Gibbs energy. This means that in the equation $\Delta G_s^* = \Delta H_s^* - T\Delta S_s^*$, the quantity $|T\Delta S_s^*|$ is much larger than $|\Delta H_s^*|$. Indeed, this is true, as can be seen in Table 3.5.

(ii) The enhancement of the structure of water (by iceberg formation or any equivalent solvent-induced effects) explains the large negative entropy of solvation. As we have seen in Section 3.7, this explanation is plausible.

(iii) Therefore, it follows from (i) and (ii) that the enhancement of the structure which is responsible for the large positive $-T\Delta S_s^*$ is also responsible for the large positive ΔG_s^*.

This argument is repeated very often in the literature. It sounds very logical; if structure enhancement is responsible for the large negative value of $T\Delta S_s^*$, and if this term is dominant in ΔG_s^*, then it follows that the structural enhancement also explains the large positive ΔG_s^*, hence explaining the hydrophobic effect.

This argument, though plausible, is incorrect. The main reason for this is that whatever the contribution of structural enhancement to $T\Delta S_s^*$, the same contribution to ΔH_s^* will occur.

This effect is referred to as the entropy-enthalpy compensation.[19] Therefore, it follows that when one forms the combination $\Delta G_s^* = \Delta H_s^* - T\Delta S_s^*$, the contribution of the structural changes in the solvent cancels out. Hence, the large positive value of ΔG_s^* cannot be explained by the enhancement of the structure of water.

3.11 Application of Entropy and the Second Law to Living Systems

The most unjustifiable, unwarranted, unacceptable, outrageous — and you can add any adjective you wish — application of the concept of entropy and the Second Law is undoubtedly in the phenomenon we call life. Here, I mean life itself, not a particular process occurring in a living system.

This application is based on the misconstrued (I would even say, perverted) interpretation of entropy as a measure of disorder, on one hand, and the view that life is a process toward more order, more structure, more organization, etc., on the other hand.

Combining these two erroneous views inevitably leads us to the association of life phenomena with a *decrease* in entropy. This in turn, leads to the erroneous (perhaps meaningless) conclusion that life is a "struggle" against the Second Law. The fact is that entropy cannot be defined for any living system, and the Second Law, in its entropy formulation, does not apply to living systems.

It is difficult to trace the origin of this misconstrued conclusion. I will be grateful for any information on this provided by the reader. There is no doubt, however, that the most prominent and influential scientist who was responsible for much of the nonsensical writings in textbooks, as well as popular science books, was Schrödinger, himself.

In his famous and widely praised book *What is Life* he expressed several times the erroneous idea that the Second Law is the "natural tendency of things to go from order to disorder," and in addition: "Life seems to be orderly and lawful behavior of matter, not based exclusively on its tendency to go over from order to disorder."

From these two assertions, Schrödinger reached the most absurd conclusion to the question: "What then is the precious something contained in our food which keeps us from death?"

His answer: "What an organism feeds upon is negative entropy ... the essential thing in metabolism is that the organism succeeds in freeing itself from all the entropy it cannot help producing while alive."

Thus, Schrödinger not only adopted the misinterpretation of entropy as a measure of disorder, and not only expressed the misconception regarding the role of entropy in living systems, but also "invented" a new concept of "negative entropy" to explain how a living system "keeps aloof of death."

It is difficult to assess the extent of the negative impact of the "negative entropy" on science and on scientists, on writers and readers of popular science books. The "negative entropy" has been transformed by Brillouin (1962) into "negentropy," and negentropy into information. Since information is "everything" (as encapsulated by Wheeler's slogan "it from bit"), then one can and has the license to say anything one wants about entropy and life — it is foolproof; no one can prove you wrong. Serious scientists followed the lead of the master, and spewed statements ranging from "life is a constant struggle against the Second Law" to the assertion that "entropy *explains* life itself." More details on this topic are available in Ben-Naim (2015a).

We emphasize again that entropy is not definable for a living system. Any statement about the entropy change in a living system is therefore meaningless. This is *a fortiori* true when we use the meaningless "negative entropy" in connection with living systems.

Here is an example of an abuse of the concept of entropy. In his book *Genetic Entropy and the Mystery of the Genome* Sanford (2005) writes:

> For decades biologists have argued on a philosophical level that the very special qualities of natural selection can easily reverse the biological effects on the Second Law of Thermodynamics. In this way, it has been argued, the degenerative effects of entropy in living systems can be negated — making life itself potentially immortal.

However, all of the analyses of this book contradict that philosophical assumption. Mutational **entropy** appears to be so strong within large genomes that selection cannot reverse it. This makes eventual extinction of such genomes inevitable. I have termed this fundamental problem **Genetic Entropy**. Genetic Entropy is not a starting axiomatic position — rather, it is a logical conclusion derived from careful analysis of how selection really operates.

Obviously, the author has no idea what entropy means, yet he uses this term in the title of the book. In most of the book, neither entropy nor the Second Law is mentioned. Only toward the end of the book do we find the above quoted paragraph, which at best can be described as pure nonsense. A more detailed review of this book may be found in Ben-Naim (2015a).

In the rest of this section, I will present a few quotations from the literature. I will leave it to the reader to decide the merits of these statements.

Here is an example from Katchalsky (1963):

Life is a constant struggle against the tendency to produce entropy by an irreversible process. The synthesis of large and information-rich macromolecules... all these are powerful antientropic forces... living organisms choose the least evil.

Volkenstein (2009):

At least we understand that life is not "antientropic," a word bereft of meaning. On the contrary, life exists because there is entropy, the export of which supports biological processes...

Hoffmann (2012):

Life uses a low-entropy source of energy (food or sunlight) and locally decreases entropy (created order by growing) at the cost of creating a lot of high-entropy waste energy (heat and chemical waste).

3.12 Application of Entropy and the Second Law to the Entire Universe

In the previous section, I admitted that I do not know the origin of the misuse of entropy in connection with life. In this section, I will describe another most common misuse of entropy and the Second Law in connection with the entire universe. In this case, I can easily pinpoint the culprit for the misuse — none other than Clausius himself.

As is well known, Clausius formulated one version of the Second Law (heat flows from a hot body to a cold body). Clausius also defined the change in entropy for the transfer of a small quantity of heat into, or out from, a system at a constant temperature. Clausius' ideas constituted the basis on which the whole science of thermodynamics was built, including the most general, most powerful and useful, Second Law of Thermodynamics. For all these achievements Clausius deserved the highest scientific credit. Unfortunately, Clausius failed in *overgeneralizing* the Second Law. His well-known and well quoted statement is

> The entropy of the universe always increases.

I do not know how Clausius arrived at this formulation of the Second Law. I can only guess what has motivated him, as well as many others who followed him, to conclude that the entropy of the universe always increases.

Consider a well-defined thermodynamic system at equilibrium. We assume that the system is large enough, so that when a small quantity of heat is transferred into a system, its temperature does not change. Therefore, according to Clausius, $dS = dQ/T$. If we define a heat transfer into the system as positive, then the corresponding change in the entropy of the *system* is also positive (see Figure 3.27a). On the other hand, if heat flows out of the system, then the entropy change of the system is negative (see Figure 3.27b).

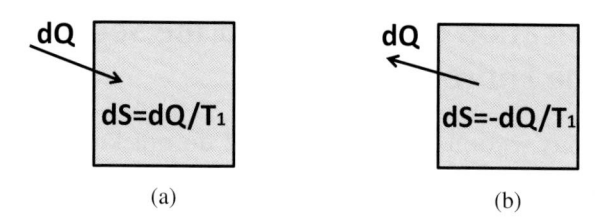

Fig. 3.27. Transfer of heat (a) into the system, and (b) out of the system.

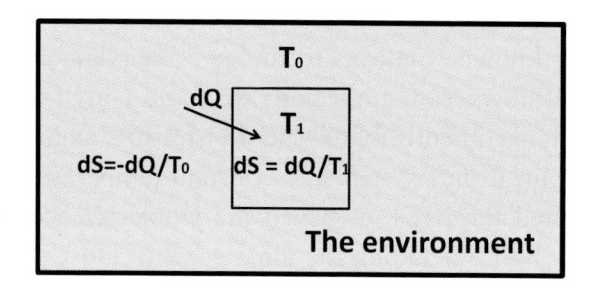

Fig. 3.28. A system in contact with an isolated heat reservoir.

Next, assume that we have the same system as in Figure 3.27, but now it is in a heat bath with a temperature higher than T_1, say $T_0 > T_1$ (see Figure 3.28).

We bring the system in contact with the heat bath briefly, and we observe that heat flows from the bath into the system. In this process the entropy of the system increases, but the entropy of the bath decreases. If the system, together with the bath, are *isolated* (this is depicted by the heavy boundaries surrounding the combined system and bath in Figure 3.28), then one can argue that the Second Law requires that in such a process the entropy of the combined system and bath must increase. Indeed, it is easy to show that $dS(system) + dS(bath)$ is positive.[20]

One can give many examples in which a spontaneous process occurred in a system at a constant temperature, for which the entropy decreases. However, the entropy of the system plus the environment (bath) must increase as a result of the spontaneous process. This is true provided the system and its environment together, form an *isolated system*.

What happens when our system is well-defined thermodynamically, say, characterized by T, V, N, and we bring the system in contact with the environment, but now the environment is the whole universe?

If the walls of the system are heat conducting, then the system will exchange heat with its surroundings, which is the entire universe (see Figure 3.29).

For simplicity, we assume that we have started with our system at equilibrium. We also assume that heat flows in or out the system very slowly, so that at each moment of time the state of the system is well defined, say, $T(t), V, N$. This means that the volume and the number of particles are unchanged, but the temperature (T) of the system changes slowly with time (t), depending on the temperature of the environment. We can also say that the system moves through a sequence of nearly equilibrium states. Therefore, at each moment of time the entropy of the system is also well defined. When the temperature of the environment is larger than that of the system, heat will (slowly) flow *into* the system causing an increase in the entropy of the system. On the other hand, when the temperature of the environment becomes lower than that of the system, heat will flow out of the system, and the entropy of the system will decrease. In this case, we have

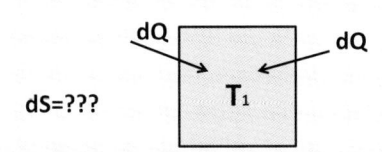

The environment is the whole universe

Fig. 3.29. A system in contact with the universe.

a spontaneous process for which the entropy of the system decreases. Does this violate the Second Law of Thermodynamics?

Of course, not! The system is not isolated, and therefore the Second Law in its entropy formulation does not apply. Also, if we take the system with its environment, here the total universe, we do not know if the whole universe is an isolated system.

It is at this point that one extrapolates from the experiment in Figure 3.28, where the system and the bath are isolated, to the experiment in Figure 3.29 where the system and its environment consist of the entire universe. In the experiment in Figure 3.28, we know how to calculate both dS(system) and dS(bath).[20] Unfortunately, in the experiment in Figure 3.29, we do not know how to calculate the entropy change of the universe!

The generalization, originally made by Clausius, is that the entropy of the *entire* universe must increase in this process. Unfortunately, such a generalization is unwarranted. Although most authors will tell you that the entropy of the universe always increases, the truth is that no one has ever measured or calculated the entropy of the universe. In fact, no one has ever *defined* the entropy of the universe. Therefore, any statement regarding the change in the "entropy of the universe" is meaningless.

Here is a quotation from a relatively recent book by Atkins (2007):

> The entropy of the universe increases in the course of any spontaneous change. The key word here is **universe**; it means, as always in thermodynamics, the system together with its surroundings. There is no prohibition of the system or the surroundings **individually** undergoing a decrease in entropy provided that there is a compensating change elsewhere.

Such a generalization is not only untrue, it is simply meaningless. This is only one example of an unwarranted generalization. Penrose (1989) and others not only discuss the entropy of the universe, but also give numbers, stating what the entropy of the universe is at present, how much it was in the early universe, and perhaps also at the Big Bang.

These are meaningless numbers assigned to a meaningless quantity (the entropy of the universe), to the state of the universe at a highly speculative time (at the Big Bang). More on this in Ben-Naim (2015a, 2016).

3.13 The Association of Entropy with the Arrow of Time

The association of entropy with the Arrow of Time has been discussed in great detail in my book *The Briefest History of Time*, Ben-Naim (2016a). Here, I will make only a few comments.

First, the *association* of entropy with time does not belong to this chapter, nor to the previous chapter. It cannot be said that this is a misuse (or abuse) of entropy. It also cannot be viewed as a misinterpretation of entropy. The best I can say is that this is a misconstrued association of entropy with time.

The origin of this association is unclear. It probably started soon after the "realization" that entropy "always increases." Since time also "always increases," it is easy to draw the conclusion that entropy must be associated with time.

There are two very well-known quotations from Eddington's (1928) book, *The Nature of the Physical World*. The first concerns the role of entropy and the Second Law, and the second introduces the idea of "time's arrow."

> The law that entropy always increases, holds, I think, the supreme position among the laws of Nature. If someone points out that your pet theory of the universe is in disagreement with Maxwell's equations — then so much the worse for Maxwell's equations. If it is found to be contradicted by observation — well, these experimentalists do bungle things sometimes. But if your theory is found to be against the second law of thermodynamics I can

give you no hope; there is nothing for it but to collapse in deep humiliation.

   Let us draw an arrow arbitrarily. If as we follow the arrow we find more and more of the random element in the state of the world, then the arrow is pointing towards the future; if the random element decreases the arrow points towards the past. That is the only distinction known to physics. This follows at once if our fundamental contention is admitted that the introduction of randomness is the only thing which cannot be undone. I shall use the phrase 'time's arrow' to express this one-way property of time which has no analogue in space.

In the first quotation Eddington reiterates the unfounded idea that "entropy always increases." Although I agree that the Second Law of Thermodynamics is unique compared with other laws of physics [see also Ben-Naim (2008, 2015a)], I do not agree with the statement that "entropy always increases."

Although it is not explicitly stated, the second quotation alludes to the connection between the Second Law and the Arrow of Time. This is clear from the association of the "random element in the state of the world" with the "arrow pointing towards the future."

In my view it is far from clear that an Arrow of Time exists. It is also clear that entropy is not associated with randomness, and it is far from clear that entropy always increases. Therefore, my conclusion is that entropy has nothing to do with time!

Of course, I am well aware of many statements in the literature which identify entropy with time. I have written about this topic in my book [see Ben-Naim (2016a)]. I will conclude this section with a quotation from Rifkin's book (1980):

> … the second law. It is the irreversible process of dissipation of energy in the world. What does it mean to say, "The world is running out of time"? Simply this: we experience the passage of time by the succession of one event after another. And every time an event occurs anywhere in this world energy is expended and the overall entropy

is increased. To say the world is running out of time then, is to say the world is running out of usable energy. In the words of Sir Arthur Eddington, "Entropy is time's arrow."

I will leave it to the reader to examine each of the quoted sentences and decide whether or not they are relevant to entropy and the Second Law.

3.14 Conclusion

In this chapter we showed several examples of the uses of entropy, the Gibbs energy and the Second Law. As I have discussed in the conclusion in Chapter 1, the entropy formulation of the Second Law is applicable only to isolated systems. The more useful application is the Second Law in its Gibbs energy formulation. We also saw a few examples of the interpretive power of entropy. In the last sections, I have briefly mentioned a few misuses of the concept of entropy and the Second Law which are very common in popular science books. In fact, most of these books discuss examples of misuses of entropy far more often than examples of proper uses.

Test Yourself After Reading This Book

The tests before and after reading this book were designed to check your preconceptions about information, the SMI, and entropy. I hope that reading this book has changed your perceptions about information, the SMI, and entropy.

Read again the questions listed in the Test Yourself Before Reading This Book section. Answer the same questions, and compare your answers. If you are interested in my answers see the final note in the Notes section of the book.

Final Test

After reading and answering all the questions in the test, before and after reading the book, you might get a sense that you have grasped what the entropy is all about, and what the Second Law means.

As it has often occurred to me on numerous occasions, when I used to teach courses on thermodynamics or statistical mechanics, I always thought that a fuzzy line ran between understanding, and a *delusion* of understanding. The best way to find out whether you really understand something, or whether you have only a delusion of understanding a concept, is to try to explain it to someone else.

Suppose that while reading this book, your young daughter curiously asks, "Daddy, what are you reading about?" Your immediate response would be "It is a book about entropy." With a puzzled look, coming from someone who has heard the word "entropy" for the first time, she asks you, "What is entropy?"

Caught with your pants down, you tell her to please wait for a few days until you finish reading the book, and then you will explain to her what entropy is all about.

Now that you have finished reading the book, you have to be ready to answer your daughter's question. This is not an easy task, and I urge you to write down your answers for your own sake; whether or not your daughter asks you. I am confident that in doing so, you will be rewarded, even if initially you might have discovered that you only had a delusion of understanding what you read in the book.

Here are some questions you might want to consider before attempting to explain entropy to your daughter:

1. What is entropy?
2. Why does entropy always increase?
3. Why does the entropy of an isolated system always increase in a spontaneous process?
4. You are given a well-defined system; either isolated, or one being at constant temperature, pressure, and number of particles. The system is initially in a constrained equilibrium state. You remove the constraints.

   a. What will happen?
   b. Why will it happen?
   c. Will the entropy increase?

I suggest that you write down your answers before you continue reading my answers.

First, I suggest that you should not tell her about Clausius' definition ($dS = dQ/T$) or Boltzmann's definition ($S = k \log W$), for this will immediately turn her off. Instead, I suggest that you start by playing with her some of the 20Q games. But before that you should explain to her and to yourself that entropy is defined for a well-defined macroscopic system at equilibrium. It is meaningless to talk about the "tendency of entropy to increase," without specifying the system.

Explain briefly what a "well-defined macroscopic system at equilibrium" means.

A macroscopic system is one with many particles of the order of 10^{23}. This is a huge number: 100,000,000,000,000,000,000,000. "Well-defined" means that it is characterized by a small number of thermodynamic parameters, say $T, P, N_1, N_2 \cdots$ (if there are external fields you need to add more parameters). Thus, a well-defined system must be a system at equilibrium. Next, you must explain that a macroscopic system has a very large number of microstates. For a classical system, think about all possible locations and velocities of all the particles. Let us call all these possibilities W. These possibilities must be consistent with the specification of the system, say E, V, N, or T, V, N. This means that all the locations must be within the region of volume V, and the total energy is E.

You can stop now and tell her that the entropy of the system is proportional to $\ln W$. Such an answer will be correct for some systems, but not the correct answer, in general. Tell her, that in more general cases, each configuration has a different probability, i.e. not every possible configuration has the same likelihood of occurrence.

If she does not understand this, give her a few simple examples of probabilities using coins or dice, and play with her the 20Q game of both uniform, and non-uniform distributions.

Once she masters the 20Q game, explain to her that people who play such games can ask either dumb questions, or smart questions. The best way to explain the different "strategies" of asking questions is to suggest to her to play a few games. She has to pay one dollar for each question, and get a big amount of money as her prize if she finds the object or the person you have chosen (do not specify the prize, as this depends on the number of possibilities).

Now she will understand that in order to win the prize by spending a minimum amount of money, she must find the object or the person, in the fewest questions possible. At this point, you might want to introduce the idea of a measure of information. Clearly, in the 20Q game you know the person you chose (or the object), while she does not have that information. By asking questions she steadily gains information from each answer you give her. She understands that her task is to gain the information on the person you chose by asking a minimum number of questions.

Once she masters the idea of gaining information for any 20Q game, tell her that a smart scientist named Shannon, found a very remarkable mathematical formula. For any game (or any experiment) with W outcomes, and given the probabilities of the outcomes, Shannon's formula gives you the minimal number of questions one needs to ask, on average, when one plays the same game many times.

The existence of such a formula is not trivial. But I am sure she would be impressed that for any 20Q game, there exists such a number, which is the minimal number of questions.

Let her digest this idea for a while before you apply it to a "well-defined system at equilibrium."

Now tell her to imagine (only to imagine) playing the 20Q game on all the microstates of a system at equilibrium. The number of possibilities is huge, and the probability of the microstates, she does not know. However, if she had grasped the *idea* (not the precise formula) of Shannon's formula, tell her that with such a huge number of microstates, it will take many ages of the universe to play the 20Q game on all the microstates. Yet, with Shannon's formula, one can *calculate* the minimal number of questions one needs to ask *if* she would have played this game.

This number is proportional to the entropy of the system.

You might want to add that this number does not depend on who is playing the game. It is a "built-in" number associated with each game. Go back to the coins and dice. Show her how the number of questions increases when the number of possibilities increases. Show her how the number of questions change when the distribution changes. Convince her that for each well-defined game there is a number which one can calculate by Shannon's formula. Specifically, for a system with many particles which is well defined thermodynamically, and at equilibrium, that number is proportional to the entropy. This answers question number 1. Now, let us go back to the questions posed in the beginning of this section

The entropy, in itself, does not increase or decrease. You must specify which system you are referring to.

For an isolated system, it is true that by removing a constraint the entropy increases. The reason is not that entropy tends to increase, but that the system always evolves from one equilibrium state to a new equilibrium state because the probability of the new state is almost one! The positive change in entropy is related to the logarithm of the ratio of the probabilities of the two states (this is not an exact answer in general, but for an isolated system the number of microstates change from $W(initial)$ to $W(final)$, and these two numbers determine the change in entropy).

Here are the answers to question 4.

The answer to (a) is simple. The system will evolve from the initial equilibrium state to a new, final equilibrium state. The answer to (b) is also simple for each of the systems $(E, V, N), (T, V, N)$ or (T, P, N), the reason *why* the system evolves to the new equilibrium state is probabilistic, i.e. the probability of a new state is overwhelmingly larger than that of the initial state.

The answer to (c) depends on the system in which the process occurred. In an isolated system (E, V, N), the entropy will increase. In the (T, V, N) system the Helmholtz energy will decrease. In the (T, P, N) system the Gibbs energy will decrease.

In summary, Shannon found that for any 20Q game, there exists a connection between the minimal number of questions you need to ask and the probabilities of all the possible objects, or outcomes, of an experiment.

When you apply the same formula for the huge number of microstates (i.e. detailed description of all the locations and velocities of all particles) of a thermodynamic system at equilibrium, you get the entropy of that system.

This entropy of a thermodynamic system at equilibrium does not change in time, and does not have a tendency to increase. It is a fixed number for any well-defined thermodynamic system at equilibrium.

If we remove a constraint, say, a partition between two compartments, or add a catalyst, then the system will evolve to a new equilibrium state. The reason is that the new state has an overwhelmingly larger probability compared to the state of the system right after removing the constraint. It turns out that in an isolated system, this change in the state is also associated with change in entropy.

Notes

Notes to the Preface
Note 1

Here is a paragraph taken from Amazon.com describing the book by Kafri and Kafri (2013).

> Why do we want more and more money regardless of how much we already have? Why do we hate to be manipulated and to lose? Why do twenty percent of the people own eighty percent of the wealth? Why in most languages, the most common word appears twice as often as the second most common word? Why the digit "1" appears in company balance sheets six and a half times more often than the digit "9"? Why does nature hate "bubbles"?
>
> The cause for all these phenomena is the very same law that makes water flow from high to low, and heat — from hot place to a cold one. This law, which for historical reasons is called the second law of thermodynamics, states that there is a never-decreasing and always increasing quantity called "entropy." The entropy represent the uncertainty of a system in hypothetical equilibrium state in which everybody and everything have equal opportunities but slim chances to win; or in other words — the majority have little and a few have a lot.

The book describes the historical evolution of the understanding of entropy, alongside the biographies of the scientists who contributed to its definition and exploration of its effects in exact sciences, communication theory, economy and sociology. The book is of interest to wide audience of scientist, engineers, students and the general public alike.

I agree only with a half of one sentence of this paragraph: "Entropy represents the uncertainty of a system." This is true provided one specifies "uncertainty" with respect to what. This topic is discussed in Chapter 1, where I will explain the uncertainty interpretation of Shannon's measure of information.

The second part of this sentence, "in which everybody and everything have equal opportunities but slim chances to win," is a meaningless sentence which has nothing to do with entropy.

Most of the book discusses many other interesting topics from "Benford's Law" to "Zipf's Law" to "Pareto Principle," and all of these have nothing to do with either entropy or the Second Law. In addition, the authors claim that the tendency of entropy to increase is the powerful machine that drives everything, including our decisions to do things, to reproduce, and even social changes associated with our culture. All these topics do not belong to physical science and certainly not to the Second Law.

Regarding the "scientific part," the authors start with a definition, or rather, an *ill definition*, of entropy. None of the phenomena discussed in their book follow from entropy, or from what they define as entropy. I will further comment on this in Chapter 1.

There are many other "inaccuracies" which are all over the book. I will mention only one here: in Figure 3 in the book the authors confuse the probability distributions of the absolute speeds of the particles with the normal distribution of velocities in one dimension. They claim that the normal distribution is an approximate to the Maxwell-Boltzmann distribution. This is not true.

Note 2

Mayhew's book's main claim is that entropy and the Second Law are superfluous, and thermodynamics would be much simpler without them.

Here is a quotation from Amazon.com describing the book:

> This book presents a completely new thermodynamic perspective, one that no longer relies upon either entropy, or the second law. Herein, we clearly demonstrate that although empirical data can disprove a given theory, it certainly does not necessarily prove any one theory, i.e. there is often more than one theory that can explain a given set of empirically known results. "Changing our perspective Part 1: A new thermodynamics" is a continuation of the revolutionary ideas presented by this author, in his several controversial peer reviewed papers. Importantly, a series of simple arguments are strung together enlightening the reader to the possibility that traditional thermodynamics has complicated what should have been, and certainly will become a simple constructive science that the majority can actually envision, comprehend, and then explain. This is not just a tweaking of the science as is done by countless other texts. It is a radical overhaul providing new insights into latent heat, lost work, kinetic theory, heat capacity, Joule's experiment, thermal radiation, Henry's law and the Clausius-Clapeyron equation. And the list just scratches the surface. Accordingly, it will up to you to decide whether you prefer the overly complex traditional conceptualizations, or a simple eloquent science. I.e. you will be asked to apply the principles of Occam's razor. All in all, this book is very insightful; bringing a whole new perspective that can be understood by most who have a scientific and/or engineering background. Importantly, this book will shock you, making you question your indoctrination in a science that you previously wrongfully believed to be mature and unrelenting.

I agree that thermodynamics would be simpler and easier without entropy and the Second Law. It would also be anemic. There are many important topics which could not be treated without the Second Law. A few examples are the Clausius-Clapeyron equation, chemical equilibrium, and phase transition. These will be discussed in Chapter 3.

Note 3

I would like to thank Oded Kafri and Kent Mayhew for giving me their books, and for many exchanges of ideas.

While reading these two books, I exchanged a few emails with Oded and Kent where I expressed my disagreement with what they wrote.

Interestingly, both Oded and Kent wrote back and told me that if only I would read their books carefully and with an open mind, I would change my mind. I have read carefully both books, and I still disagree with both of them. The authors ascribed my insistence on disagreeing with them to a result of indoctrination and brainwashing by the common views of thermodynamics — and if only I would open my mind, I would see the light.

I admit that the views expressed by both of them are non-conventional, and perhaps also original. However, I believe that my disagreement with them is not because I am indoctrinated and brainwashed by the commonly accepted views, but simply because I feel that their views are fundamentally wrong.

I might add here that there are many good textbooks on thermodynamics which present both the definition of entropy and its application adequately, but fail only in the interpretation of entropy.

The books of Kafri and Mayhew are exceptional in the sense that they are both wrong in their entirety: in their definitions, interpretations, *and* applications of entropy.

As I explained in Ben-Naim (2015a), my views are not a result of being "brainwashed" by the accepted views, but rather a result of washing my brain off from accepted views and developing my own, original, and new approach.

Oded Kafri asked me to add the following two notes:

1. Kafri and Kafri (2013) claim that in every irreversible energy flow in the universe, entropy increases. Therefore the driving force of everything is the Second Law.
2. In this book Ben-Naim attacks both the formulation and the interpretation of entropy in our book *Entropy: God's Dice Game*. Ben-Naim (BN) was kind enough to let me defend our view.

The main difference between BN's view and our view lies in BN's erroneous formulation of Clausius entropy which unfortunately appears in several books, especially in physical chemistry, namely,

$$dS = \frac{dQ}{T}. \qquad (1)$$

The right expression of entropy is

$$S = \frac{Q}{T}, \qquad (2)$$

namely, the change in entropy S is the energy removed or added Q divided by the temperature T.

Equations (1) and (2) look identical but they are not! Q is heat. Heat is a transient quantity similar to work. One cannot store work in a box. Similarly, heat is energy E added or removed from a bath. Therefore, the right differential representation of entropy is $dS = \left(\frac{dE}{T}\right)_{rev}$ which has in many cases a completely different meaning.

For example, BN's expression (1) yields a paradox according to which Clausius's famous law (the Second Law), that the entropy in the universe is always increasing, is erroneous. However, in expression (2) S is the change of entropy as a result of removing the energy Q. This energy should be dumped somewhere (or vice versa) such that the total change of entropy is $\frac{Q}{T_1} - \frac{Q}{T_2} \geq 0$. This is Carnot's finding and therefore Clausius' famous sentence that the entropy of the universe is always increasing is always true, and remains one of the most important laws of nature.

My explanations above are not new by any means. However, in our book we draw the attention to the fact that Shannon entropy is also a transient quantity similarly to heat. The hot reservoir is Bob sending an energetic file to Alice, the cold reservoir. If the entropy of the bit "one" is k_B namely, one Boltzmann constant (the entropy of a hot harmonic oscillator), then Shannon entropy is equal to Boltzmann-Gibbs entropy.

BN does not like the interpretation of entropy as uncertainty. He asks: "uncertainty of what?" The answer is understandable from the transient nature of entropy. Let us look at the most common example of ideal monoatomic gas. What is the energy of N molecules of gas?

The common answer is that the energy is that of the translational degrees of freedom, namely, 1.5 $k_B NT$. This answer is true only if we neglect the huge amount of energy of the atoms (electrons and nucleuses) themselves. These kinds of separation is always done in physics. Therefore, if I remove Q energy from monoatomic ideal gas, the uncertainty of the gas is the change of the translational uncertainty due to the cooling of the atoms. Similarly, when Bob sends Alice a file, the uncertainty is the logarithm of number of contents possible. N bits of file means 2^N possible contents, and $S = k_B N \ln 2$ entropy.

Needless to say, I disagree with everything said in these two notes, by Kafri, as well as in their book. The reasons are explained in Chapter 1. Of course one can *define* a quantity S by: $S = Q/T$. Unfortunately, this quantity has nothing to do with the entropy of the system, nor has it any of the properties of the entropy defined for a system at equilibrium. This renders the Kafris' book empty of any scientific content.

Notes to Chapter 1
Note 1

Denoting by $\Delta Q(Hot)$ and $\Delta Q(Cold)$ the amounts of heat flowing inside and outside the engine, respectively. The efficiency of the heat engine is defined by

$$\eta = \frac{\Delta W}{\Delta Q(Hot)} = \frac{\Delta Q(Hot) - \Delta Q(Cold)}{\Delta Q(Hot)} = 1 - \frac{\Delta Q(Cold)}{\Delta Q(Hot)}$$

Where ΔW denotes the amount of useful work done by the system (e.g. lifting a weight).

In the Carnot engine operating between the two temperatures T_2 and T_1 ($T_2 > T_1$), the efficiency is given by

$$\eta = \frac{T_2 - T_1}{T_2} = 1 - \frac{T_1}{T_2} \leq 1$$

Since $T_1 < T_2$, $\eta < 1$. The efficiency is zero when $T_1 = T_2$, and it approaches one when $T_1/T_2 \to 0$.

Note 2

Note that this is a *specific process of mixing* occurring in an isolated system. There is a mixing process with no change of entropy, and also there is a process of *demixing* with *positive* change of entropy (i.e. a spontaneous demixing process in an isolated system — see Section 2.4).

Note 3

This is true when the combined system of the two bodies is isolated.

Note 4

Sometimes you see a slightly different notation, i.e. dQ_{rev}/T, where "rev" is short for reversible. We will not need this notion here. We require that the system is large enough at a specific temperature T, and that dQ is small enough so that it does not change the temperature of the system. Some authors use the notation δQ to emphasize that this quantity is not an exact differential. On the other hand, dS is an exact differential. This means that there exists a function S, which is a state-function, i.e. a function of the parameters describing the system, say T, P, N, and is differentiable with respect to these variables.

If one defines $S = Q/T$, as in Kafri and Kafri (2013), then either Q becomes a state function (which it is not!) or S becomes a quantity which, like heat, flows (which it does not!). And besides, with such a "definition" one cannot calculate the entropy of any system, nor entropy changes of any process.

Note 5

Clausius (1867) quoted by Cooper (1968).

Note 6

This is discussed in detail in Ben-Naim (2008).

Note 7

Battino (2007).

Note 8

We shall use the term *quasi-static* instead of reversible. The terms *reversible* and *irreversible* are loaded with different meanings; see Ben-Naim (2011a, 2016a).

Note 9

The work of expansion from V_1 to V_2 at constant pressure. P is calculated from

$$\Delta W = -\int_{V_1}^{V_2} P dV = -P(V_2 - V_1)$$

The negative sign of ΔW means that work is done *by* the system. In the process depicted in Figure 1.2a the pressure is *not constant*, and in fact it is not even defined along the way from the initial to the final state. The trick is to do the process along a large sequence of small steps. See Figures 1.3b and 1.3c.

In the limit of very small steps we can calculate the work as

$$\Delta W = -\int_{V_1}^{V_2} P dV = -Nk_B T \int_{V_1}^{V_2} \frac{dV}{V} = -Nk_B T \ln 2$$

And the corresponding entropy change is calculated from the equation $TdS + dW = TdS - PdV = 0$.

Hence,

$$\Delta S = -\int_{V_1}^{V_2} \frac{PdV}{T} = Nk_B \int_{V_1}^{V_2} \frac{dV}{V} = Nk_B \ln 2$$

It should be noted that Mayhew (2015b) considered the expansion process from the initial volume V_1 to the final volume V_3. He correctly quoted the change in the entropy for such an expansion process (presumably of an ideal gas, but this is not mentioned):

$$\Delta S = k \ln \left(\frac{V_3}{V_1} \right)$$

Then surprisingly, he claims that since this equation does not take into account the increase in volume $(V_3 - V_1)$, it is "unrealistic." Unfortunately, V_3 includes the "increase of volume," and the above change in entropy is the correct one for the expansion of an ideal gas from V_1 to V_3.

Note 10

Boltzmann (1896). In fact, the entropy does not change with time at all. We shall discuss this in Section 3.13.

Note 11

Ben-Naim (2007).

Note 12

The Boltzmann H-function is defined by the equation

$$H = \int f(v,t) \log f(v,t) dv$$

where $f(v,t)$ is the distribution of velocities of the particles at time t.

Boltzmann showed that this function always is a non-increasing function of time

$$\frac{dH(t)}{dt} \leq 0$$

Thus, the function $-H(t)$ is a function which is a non-decreasing function of time. This function $(-H(t))$ was erroneously identified with entropy. Clearly $-H(t)$ is an SMI, *not* entropy. When the system reaches equilibrium, the SMI of both the locations and velocities become proportional to entropy.

In my view, all these paradoxes and objections resulted from referring to the Boltzmann H-function, or rather $(-H)$ function, as entropy. However, $(-H)$ is not entropy. As we shall see in Section 1.4, the entropy of a thermodynamic system is the Shannon measure of information (SMI) with respect to both the probability distribution of locations and velocities at equilibrium. The H-function constructed by Boltzmann is built only on the probability distribution of velocities $f(v, t)$ at time t, and this is *not* the distribution at equilibrium.

Boltzmann also showed that at equilibrium the probability distribution of velocities attends the Maxwell-Boltzmann distribution. Thus, the H-function is an SMI, not entropy. This function was shown to be a decreasing function of time, or equivalently $(-H)$ was shown as an increasing function of time. As such, it is not entropy. When we reach *equilibrium*, then $f(v, t)$ will attain the Maxwell-Boltzmann distribution. Add to this also the equilibrium distribution, with respect to locations, and you get the entropy. This quantity is the maximal value of the SMI, based on *both* the locational and velocity distribution at equilibrium. As such, the entropy is *not* a function of time.

Thus, all of the objections to Boltzmann's entropy were misdirected. We shall discuss the SMI and its maximal value in Section 1.4.5.

Note 13

In mathematics, the notion of probability distribution is sometimes defined differently. Also, there is a different definition for the case of a discrete set of outcomes (such as the results of throwing a die), and a continuous set (at which point on a board a dart hits). We shall not need the continuous case, nor the case of infinite number of outcomes.

Note 14

The function drawn in Figure 1.5 is the SMI for the case of two outcomes, H and T. Since $p_H + p_T = 1$, it follows that the SMI is a function of only one variable $p = p_H$ (Note that $(1 - p) = p_T$). Thus for this case we have the formula

$$H(p) = -p \log p - (1 - p) \log(1 - p)$$

Not also that the limit of $p \log p$, when p tends to zero, is zero.

Note 15

The discussion in this example is very qualitative. However, it is qualitatively clear, that I gave you maximum information in case (a). This means you can "guess" with certainty the outcome. You got quite a lot of information in case (b). This means that you can guess with almost certainty that the outcome is H. In case (c), the information you got is minimal. The knowledge of the distribution in this case does not give you any clue, or any hint how to make an "intelligent" guess.

Note 16

Intuitively, it is clear that case (a) is the easiest to play. Therefore I will choose this game. Since I know which result occurred, I do not need to ask any questions.

The last case (c), is the most "difficult" to play. It is difficult, in the sense that the information given on the distribution does not provide us with any clues about the outcomes. In this case, I must ask, on average, about 2.58 questions. First question: Was the outcome between 1 and 3? Assuming the answer is "Yes," I will continue with the next question, "Is it 1?" If the answer is "Yes," then I end the game and get the prize. If the answer is "No," then I have to ask one more question to know the answer.

The case (b) is intermediate between (a) and (b). The best strategy to ask questions is to first ask, "Is it 1?" There is a large probability (0.8) that you will get a "Yes" answer, hence ending the game. If the answer is "No," then I need to ask one more question to determine which outcome (2 or 3) had occurred. Thus, in this case I will need to ask between one and two questions. For more precise answers on the average number of questions, we need to calculate the SMI for this case, which is

$$H = -(0.8)\log(0.8) - 2(0.1)\log(0.1) = 0.92$$

This means that, on average we shall need about one question.

Notes | 245

Note 17

The relationship between the absolute temperature T and the average kinetic energy E_k of the particles is given by: $k_B T = \frac{2}{3}\left[\frac{m\langle v^2\rangle}{2}\right] = \frac{2}{3}E_k$.

Note 18

We impose here a specification of the configuration for each particle within a small but finite-sized "box." Later, we shall see that quantum mechanics imposes on us the size of the "box."

Note 19

For details see Ben-Naim (2008, 2012).
The 1-D distribution of velocities is the normal distribution. It is also referred to as the Maxwell-Boltzmann distribution. This is an *exact* normal distribution. Sometimes, this 1-D distribution is confused with the distribution of the speeds of particles, i.e. the absolute value of the velocity in 3-D which is defined by $v = \sqrt{v_x^2 + v_y^2 + v_z^2}$.

Note 20

A quick clarification. A distribution is a vector of numbers $p_1, \ldots, p_n$ which are probabilities, i.e. p_i is the probability of the event i. The "probability" mentioned in the text is $\Pr(p_1, \ldots, p_n)$, i.e. the probability (Pr) of obtaining the distribution $(p_1, \ldots, p_n)$.

Note 21

In quantum mechanics, the microscopic states are defined as the solutions of the so-called stationary Schrödinger equation. The classical microscopic state is easier to visualize, i.e. specifying all the locations and velocities of the particles. The quantum mechanical microstates are, in general, not easy to visualize.

Note 22

See any book on probability, or Ben-Naim (2015b). Basically it says that if we perform a large number of experiments, the relative frequency of an event will approach the probability of that event.

Note 23

The number of specific configurations consistent with n particles in L and $N - n$ in R is

$$W(n) = \frac{N!}{n!(N-n)!}.$$

The corresponding probability is:

$$P_N(n) = \frac{W(n)}{W_T} = \left(\frac{1}{2}\right)^N \frac{N!}{n!(N-n)!},$$

where W_T is the total number of specific configurations given by

$$W_T = \sum_{n=0}^{N} W(n) = \sum_{n=0}^{N} \frac{N!}{n!(N-n)!} = 2^N.$$

Note 24

The classical definition of probability applies to an experiment for which all the elementary events have equal probabilities. In our case all the $W(n)$ specific events have equal probabilities. See also Ben-Naim (2015a).

Note 25

For any distribution (p, q) we can define the SMI as

$$\text{SMI}(p, q) = -p \log p - q \log q = (-p \ln p - q \ln q)/\ln 2$$

(Note that we used both the natural logarithm ln, and the logarithm to the base 2, $\log_2(\ )$.

This is the SMI per particle in the system. It measures the uncertainty in the location of each particle, with respect to being in L or in R. When $N \to \infty$, we can use the Stirling approximation to rewrite the logarithm of the number of states W_N as

$$\ln W(p,q) = \ln \binom{N}{pN} \approx -N[p \ln p + q \ln q] - \frac{1}{2}\ln(2\pi Npq)$$

$$= N \times \text{SMI}(p,q) \ln 2 - \frac{1}{2}\ln(2\pi Npq), \quad \text{for large } N$$

Hence, in this approximation, the probability of finding the distribution $(p, 1-p)$ is

$$\Pr(p,q) \approx \left(\frac{1}{2}\right)^N \frac{\exp[N \times \text{SMI}(p,q) \ln 2]}{\sqrt{2\pi Npq}}$$

or equivalently,

$$\Pr(p,q) \approx \left(\frac{1}{2}\right)^N \frac{2^{N \times \text{SMI}(p,q)}}{\sqrt{2\pi Npq}}$$

Note 26

I know that most textbooks will tell you that although the entropy change in processes 5 or 6 might be negative, the entropy of the "entire universe" will increase. This is unfortunately not true — we cannot even define the entropy of the entire universe, let alone know the sign of its entropy change. We shall come back to this topic in Chapter 3.

Note 27

The chemical potential is the quantity which "directs the flow of matter." As heat flows from high to low temperature, and "volume" flows from low to high pressure, matter flows from high to low chemical potential. If there are different compounds in a system, then we define the chemical potential for each compound as the partial derivative of the Gibbs energy with respect to the number of particles (or moles) of that compound. The chemical potential is most useful in studying chemical reactions. In these cases the "flow" means from some reactant compounds to some products. An example is the folding of a protein. If the chemical potential of the unfolded form is higher than the folded form, then the protein will "flow" from the unfolded to the folded form. At equilibrium the chemical potentials of the two forms must be equal. See also Chapter 3.

Note 28

See Scully (2007), page 62:
Entropy not only explains the arrow of time, it also explains its existence; it is "time."
Highly purified nonsense.

Note 29

The qualitative answers are as follows:
Denote by $S(E, V, N)$ the entropy of the left-hand compartment in Figure 1.28a.

1. The entropy of the system in Figure 1.28a is $S(E_1, V, N_1) + S(E_2, V, N_2)$.
2. The entropy of the system in Figure 1.28b is $S(E_1 + E_2, 2V, N_1 + N_2)$. This, in general, is different from the entropy of Figure 1.28a.
3. The probability is zero; the distribution of particles is fixed: $x_1 = N_1/N$ and $x_2 = N_2/N$
4. In 1.28b, one can find any distribution of particles with finite probability $\Pr(x_1, x_2)$; see Box 1.5.

Note 30

See http://entropysite.oxy.edu.

Note 31

Here are some quotations from Kafri and Kafri (2013), which I believe are meaningless:

> Entropy is always greater than Q/T, unless the system is in equilibrium, in which case they are equal. This inequality, now called "Clausius's inequality," is the formal expression of the second law.

Entropy is a quantity defined for a system *at equilibrium*. Q is a quantity of heat transferred into or out from a system. Therefore the above quotation is meaningless.

> This means that any action in nature — whether "natural" or man-made — increases entropy. In other words, entropy's tendency to increase is the tendency of nature, including us humans, to make energy flow.

This is a typical meaningless statement, appearing in many popular science books.

> Clausius' inequality (the value Q/T is smaller than entropy) occurs in two instances: when a process is irreversible; when a system is not in equilibrium. Under such conditions, the maximum value of Q/T can be S, such that $S \geq Q/T$, or entropy is always greater than Q/T, unless the system is in equilibrium, in which case they are equal.

This is another meaningless statement. The maximum of Q/T can be any number, and it has nothing to do with entropy.

In conclusion, The "entropy" defined by Kafri, is not entropy, and all the discussions and applications of their "entropy" have nothing to do with entropy.

Notes to Chapter 2
Note 1

It seems to me, that if the absolute temperature had been defined after the development of the kinetic theory of gases, and recognized as a measure of the average kinetic energy of the particles, it would have been bestowed with the units of energy. These units are more "natural" and more appropriate units for temperature. Having temperature in units of energy would automatically render the Boltzmann constant superfluous and the Boltzmann entropy a dimensionless quantity. Such a dimensionless quantity will still be a state function. In this case, it would be easier to accept the interpretation of entropy as a measure of information. More on this aspect of entropy in Ben-Naim (2008).

Note 2

The fact that matter can be converted to energy and vice versa, invalidates the conservation principle for matter and energy, individually. Instead, we have the conservation of the total energy and matter.

Note 3

Please note that the title of Section 2.2.2 is meaningless without specifying the system, the entropy of which increases.

Note 4

I was so deeply impressed by these statements which prompted me to write to Oded Kafri to ask him whether making love and having children are also the results of the Second Law. His answer was: "Of course, everything that happens in the universe is a result of the Second Law. Read my book and you will be convinced."

I read the book and I was convinced that the authors are totally wrong. Besides, "waterfalls" and "heat flows" are not governed by the same law! In the book's prologue, the authors write: "Entropy's tendency to increase is the source of all changes in our universe."

It is ironic that throughout their book, Kafri and Kafri (2013) repeatedly remind the reader that entropy is not well understood, and that this fact allows people *to say anything that come to their minds about entropy*. This is painfully true, and this is exactly what these authors do throughout the whole book.

Note 5

As we have discussed in Section 1.3.3, the conflict between the reversibility of the equations of motion and the apparent irreversibility of thermodynamics arises mainly from the H-theorem, and referring to the $-H(t)$ as entropy. However, $-H(t)$ is *not* entropy for two reasons: First, it is only the SMI defined on the velocity distribution $f(v,t)$, and not on the distribution of both the locations and velocities; second, the $-H(t)$ is a function of time, but the entropy is related to the *limit* of $-H(t)$ as $t \to \infty$, i.e. at equilibrium. A summary of attempts to find the dynamic origin of the thermodynamic irreversibility may be found in Mackey (2003).

Note 6

The change in entropy for the process II in Figure 2.2b is calculated from

$$\Delta S = Nk_B \ln \frac{T}{T_1} + Nk_B \ln \frac{T}{T_2}$$

where k_B is the Boltzmann constant, T_1 and T_2 are the temperatures of the right and left compartments, respectively, and T is the final equilibrium temperature, which in this case is $T = \frac{T_1+T_2}{2}$. Note that the first term is positive ($T > T_1$), the second term is negative ($T < T_2$), but the sum is positive. This follows from the inequality $\frac{T_1+T_2}{2} \geq \sqrt{T_1 T_2}$.

The arithmetic average is always larger than the geometric average (unless $T_1 = T_2$).

Note 7

Boltzmann (1896).

Note 8

Some authors say that "probability tends to increase" [see for example, Brillouin (1962)]. The probability does not have a tendency to increase; it is the state of the system which changes from a state of lower probability to a state of higher probability.

Note 9

We calculated the entropy change for two ideal gases where initially we have one mole of A molecules in a volume V and temperature T, and in a different compartment we have one mole of B molecules in another volume V and temperature T. The whole system is isolated. The entropy change can be easily calculated from the generalization of the entropy function for mixtures [see Ben-Naim (2012)].

Note 10

For a quick qualitative answer we can assume that most of the ice in case (a) will evaporate, and the temperature of the gas will be insignificantly lowered. In case (b) we can assume that most of the gas will be crystalized into ice, and the temperature of the ice will be insignificantly increased.

The qualitative answers for case (a) are as follows:

1. Because of the overwhelming quantity of the steam, the ice will evaporate.
2. The final temperature will always be between the initial and final temperatures of the two systems. However, because of the larger quantity of the steam, we can assume that the temperature of the gas will slightly decrease.
3. Since this process is a spontaneous process in an isolated system, the total entropy change will increase.
4. Although the temperature of the gas will be slightly lower than its original temperature, we shall probably view the state of the entire system as being more disordered than the initial state.

Note that for any finite quantities of ice and steam, we can calculate quantitatively the final temperature as well as the entropy changes if we know the heat capacity of the steam and of the ice as a function of temperature, as well as the heat of evaporation of the ice.

The qualitative answers for case (b) are as follows:

1. Because of the overwhelming quantity of the ice, nearly all the steam will condense and form ice.
2. The final temperature will always be between the initial and final temperatures of the two systems. However, because of the larger quantity of the ice, we can assume that the temperature of the ice will slightly increase.
3. Since this process is a spontaneous process in an isolated system, the total entropy change will increase.
4. Although the temperature of the ice will be slightly higher than its original temperature, we shall probably view the state of the entire system as being more ordered than the initial state.

Note 11

Note that by true value, I mean the value you obtain by using the Third Law to calculate the absolute value of the entropy of the system, if that can be done. Alternatively, one can calculate the absolute entropy using statistical mechanics. This is in general quite a difficult calculation. For very weak interactions one can also use the SMI method. See Ben-Naim (2008, 2012).

Note 12

When the system expands, the average distance between the particles will increase. Therefore, the average interaction energy is smaller in absolute values. Since the system is isolated, the energy required to separate (on average) the particles will come from the kinetic energy of the atoms, hence the temperature will decrease. For more details, see Ben-Naim (2012).

Note 13

Note that in this expansion you need energy to increase the average intermolecular distances. This energy will come from heat which will flow *into* the system to maintain its temperature constant.

Note 14

This process was referred to as an assimilation process, i.e. a process in which only the number of indistinguishable particles is changed. In thermodynamics, we can combine the two systems on the left-hand side of Figure 2.12a, and compress the system to the volume V. The entropy change in this process is $-Nk_B \ln 2$. For more details, see Ben-Naim (2008, 2012)].

Note 15

"Freedom" does not mean "range of possible actions." There can be a large range of possibilities but no freedom of choice, and there can be freedom of choice but from a small number of possibilities.

Note 16

For more details on the various meanings of irreversibility see Ben-Naim (2011a, 2015a).

Note 17

The title of Atkins' book, *Four Laws That Drive the Universe*, is also misleading. A more appropriate title would be *Four Laws That Do Not Drive the Universe*.

Note 18

One can interpret x_l as follows: We pick up a molecule at random. The probability that it is an l molecule is x_l. The probability that it is a d molecule is x_d.

Note 19

My answers to the questions in Section 2.12 are as follows:

1. At equilibrium we still find equal amounts of l and d, i.e. the mole fractions of the two enantiomers will be $x_l = x_d = \frac{1}{2}$.
2. The change in entropy will be $\Delta S = 10R \ln 2$.
3. Starting with (a), the entropy will first increase, until we reach (e), then it will decrease. The entropy of (h) will be the same as (a).

4. I have no idea how to answer (i) to (v). The answer to (vi) is simple:

$$0 = \text{SMI}(a) < \text{SMI}(b) < \text{SMI}(c) < \text{SMI}(d) < \text{SMI}(e)$$
$$> \text{SMI}(f) > \text{SMI}(g) > \text{SMI}(h) = 0$$

Note that these values of the SMI are the changes in the SMI relative to (a).

5. The only valid correlation is between the changes in the SMI and the changes in entropy.

Notes to Chapter 3
Note 1

By theoretical value of the entropy we mean the calculated entropy based on the entropy function (Sackur-Tetrode equation for ideal gas, as well as contribution to entropy from rotational, vibrational, and other degrees of freedom). For details, see Rushbrooke (1949) and Wilks (1961).

Note 2

There are many substances for which we cannot determine the absolute value of the entropy either experimentally or theoretically. Nevertheless, we believe that one can assign an entropy value to any well-defined thermodynamic system.

Note 3

It should be noted, however, that in most textbooks the positive change in entropy is erroneously ascribed to the mixing of the two gases — lending credibility to the disorder interpretation of entropy.

This interpretation has its origin in Gibbs' writings, and is discussed in great detail in Ben-Naim (2008, 2012, 2015a).

Note 4

In 2006, the editors of *Science* listed 125 big questions of science. One of them is the following:

> How do proteins find their partners? Protein-protein interactions are at the heart of life. To understand how partners come together in precise orientations in seconds, researchers need to know more about the cell's biochemistry and structural organization.

Note 5

This is the competition between the energy and the entropy terms in the Helmholtz or the Gibbs energy. If there is no "glue" that binds the two monomers, i.e. ΔE is negligible, then the dissociation of the monomer would be favorable. Dissociation involves $\Delta S > 0$, hence $-T\Delta S < 0$, hence $\Delta G < 0$. This means that dissociation would be favored.

Note 6

Another big question in the 2006 *Science* list (see Note 4 above) is the following:

> Can we predict how proteins will fold? Out of a near infinitude of possible ways to fold, protein picks one in just tens of microseconds. The same task takes 30 years of computer time.

Note 7

This is essentially Anfinsen's experiments on folding and unfolding of proteins in aqueous solutions, Anfinsen (1973). For more details see Ben-Naim (2011, 2014, 2016b).

Note 8

In vivo, means in the living cell. There might be agents which help the protein to fold into the right structure. Here we discuss only spontaneous folding of proteins in aqueous solution.

Note 9

The Clapeyron equation is easily obtained from the condition of equilibrium between two phases, α and β, $\mu_\alpha = \mu_\beta$ where μ is the chemical potential. Taking the derivative $\mu_\alpha - \mu_\beta$ along the equilibrium line, we get

$$0 = \frac{d(\mu_\alpha - \mu_\beta)}{dT} = -\Delta S + \Delta V \left(\frac{dP}{dT}\right)_{eq}$$

After rearrangement, we get the Clapeyron equation

$$\left(\frac{dP}{dT}\right)_{eq} = \frac{\Delta S}{\Delta V}$$

This equation is exact and applies to any two phases at equilibrium. One can get an approximate equation for the case of vaporization of a liquid or a solid phase. In such a case, one assumes two approximations: (1) the molar volume of the gas V_g is much larger than the molar volume of the liquid or the solid phase, and (2) the vapor behaves as in an ideal gas. If these two assumptions are valid, then an approximate equation is obtained which is called the Clausius Clapeyron equation:

$$\frac{dP}{dT} = \frac{P \Delta H_{vap}}{RT^2}$$

From this equation one can get the dependence of the vapor pressure on the temperature.

It should be noted that Mayhew (2015a) refers to the Clausius-Clapeyron equation, the approximate relationship presented above. He also developed the relationship between the vapor pressure and temperature using kinetic arguments instead of thermodynamic arguments. Regarding the "traditional" approach, the author comments: "Again, the Clausius-Clapeyron equation suffers the same traditional detriment: $L(l \to g) \ggg RT$, for all substances." (L is the latent heat of vaporization.)

I sharply disagree with this statement. First, the Clapeyron equation is the exact equation $\frac{dP}{dT} = \frac{\Delta S}{\Delta V}$, not the approximate equation discussed by the author. Second, there is nothing detrimental in the exact Clapeyron equation. And finally, I would challenge the author to derive the exact Clapeyron equation without using entropy or the Second Law.

Note 10

The misinterpretation of the "entropy change" in this process is a result of misinterpretation of entropy as a measure of disorder. It is also a result of the fact that thermodynamics by itself cannot offer any valid molecular interpretation for the entropy change in any process.

Note 11

A challenging question to Kent Mayhew: Can you explain the entropy change in this process without entropy?

A challenging question to Oded Kafri: Can you calculate the entropy change in this process from your definition of entropy?

Note 12

Start from the identity $2^{2N} = (1+1)^{2N} = \sum_{i=0}^{2N} \binom{2N}{i} > \frac{(2N)!}{N!N!}$.

The inequality on the right-hand side is a result of taking one term out of $(2N+1)$ non-negative terms. Taking the logarithm of the two sides, we get $2N \ln 2 > \ln \frac{(2N)!}{(N!)^2}$, or equivalently, $2N \ln 2 + \ln \frac{(N!)^2}{(2N)!} > 0$.

Note 13

We use the Stirling approximation in the form $\ln N! \approx N \ln N - N$, and also, $\ln(2N)! \approx 2N \ln 2N - 2N$.
Therefore,

$$\ln \frac{(N!)^2}{(2N)!} \approx 2[N \ln N - N] - [2N \ln 2N - 2N]$$
$$= 2N \ln N - 2N \ln 2N$$
$$= 2N \ln \frac{1}{2} = -2N \ln 2$$

Note 14

Gibbs (1906).

Note 15

For reviews on this topic, see Eliel (1962):

Note 16

One can interpret x_l as follows. We pick up a molecule at random. The probability that it is an l molecule is x_l. The probability that it is a d molecule is x_d.

Note 17

The probability of obtaining the distribution (x_l, x_d) is $\Pr(x_l, x_d) = \frac{N!}{N_l! N_d!}$, where $x_l = \frac{N_l}{N}$ and $x_d = \frac{N_d}{N}$.

We use the Stirling approximation for each of the factorials to obtain

$$\ln \Pr(x_l, x_d) = \ln N! - \ln N_l! - \ln N_d!$$
$$\approx N \ln N - N_l \ln N_l - N_d \ln N_d$$

$$= N[-x_l \ln x_l - x_d \ln x_d]$$
$$= N(\ln 2) H(x_l, x_d)$$

Note 18

For dilute solutions, of say, methane in water, the solubility ρ_{sol} is related to the Gibbs energy of solvation by the equation $\Delta G_S^* = -k_B T \ln \rho_{sol}$.

Note 19

The structural enhancement by an inert solute may be viewed as a "flow" of water molecules from less structured to more structured species of water molecules. Since this "flow" occurs at equilibrium, $\Delta G(structural\ change) = 0$, therefore we have $\Delta H(structural\ change) = T\Delta S(structural\ change)$, which means an exact compensation between the ΔH and $T\Delta S$ associated with the structural changes.

Note 20

$$dS(system) = dQ/T_1, \quad dS(bath) = -dQ/T_0$$

Therefore,

$$dS(total) = dQ/T_1 - dQ/T_0 = dQ \left(\frac{T_0 - T_1}{T_1 T_0} \right)$$

Since $dQ > 0$, and $T_0 > T_1$, we have a positive change of entropy of the combined system.

Final Notes to the Book

Here are answers to the questions posed in the tests for before and after reading the book.

The Qualitative Test

Answer to (i). The information you got is exactly what is written in items (1) to (10).

Answer to (ii). How much *new* information you got depends on what you already knew. Therefore, the answers to this question is very subjective.

Answer to (iii). How surprised were you depends on what you knew. Therefore, again the answer is very subjective.

Answer to (iv). The "size" of the information depends on the method you adopted to measure the information. If you measure the size according to the number of words or letters in the message, then the "size" of (5) is slightly larger than the previous ones. On the other hand, the size of (6) is larger than (5) although both convey the same information. The "size" of (9) is larger than (10) although they convey the same information. This "size" is different from the SMI, which we asked about the quantitative tests.

The Quantitative Tests

In these tests, you have to distinguish between the information given, the size of the information contained in the distribution (or in the experiment), and the size of the information (often referred to as self-information) associated with the knowledge of the specific result of the experiment.

Answers to question (i)

The information you got is exactly what is written in the sentences (1) to (4) in all the cases. For instance, the information I got in Case A number 2, is simply "The outcome was 4."

Answers to question (ii)

Here, the average uncertainty is different for each case. In case A, the uncertainty is maximal. In case C, the uncertainty is minimal. In case B, the uncertainty is somewhere between cases A and C. In case D, the uncertainty is not known.

Note that the answers given above are relevant to the information in item (i) in each case, (A, B, and C), i.e. knowing the distribution, we can estimate the average uncertainty.

The information (2), (3), and (4) for each case does not tell us anything about the uncertainty involved in the outcomes of the dice.

Answers to question (iii)

The question is relevant to items of information (2), (3), and (4).

In case A

I am equally surprised to learn when either the result "4" or "1" were obtained. This can be measured by log 6 for any specific outcome "k." On the other hand, for the outcome "even," the surprizal is smaller, and may be measured by log 2.

In case B

The surprisal in (2) is log 4, which is larger than in (3) which is log 2. Clearly, given the distribution in (1) we are less surprised in knowing that the outcome is "1," than in knowing that the outcome is "4." The surprizal of knowing that the outcome is "even" is larger than knowing that the outcome is "odd. This is so because the probability of getting "odd" in case B is $\frac{1}{2} + \frac{1}{4}$, whereas for getting "even" it is $\frac{1}{4}$.

In case C

I am not surprised at all that the outcome is "1," but I am infinitely surprised to hear that the result is either "4," or "even."

In case D

I cannot answer this question since I do not know the distribution.

Answers to question (iv):

The SMI for the distribution in case 1 is $\log_2 6$. In case B, it is $-\frac{1}{2}\log_2\frac{1}{2} - \frac{1}{4}\log_2\frac{1}{4} - \frac{1}{4}\log_2\frac{1}{4} = \frac{3}{2}$. In case C, the SMI is zero, and in case D, I do not know the SMI.

The answers given above pertain to item (1) in each case, i.e. given the probability distribution, we can calculate the SMI.

Knowing any of the items (2), (3), or (4) in each case does not provide the SMI on the probability distribution. Note that some authors assign a measure of "self-information" to a single outcome of an experiment. Clearly, you do not get more or less information when an outcome "k" or "l" has occurred.

Regarding Table I, all the statements are either false, or unknown:

Regarding Table II, the answers depend on how you define *information*, and how you *measure* information. Without these definitions one cannot say whether the statements are either true or false.

A Final Challenge — Something to Tickle Your Grey Cells

After reading, and hopefully understanding and enjoying, the book's message, I offer you a recreational challenge which will also, hopefully, tickle your brains. I listed below a few "conjectures" on the thermodynamics of cockroaches. All of these were deduced from statements made by well-known scientists. Unfortunately, I cannot prove any of these conjectures. I am offering substantial prize money to anyone who will send me a convincing proof of any of the following conjectures:

1. Cockroaches are considered "social" insects. They tend to aggregate and some males practice courtship rituals in accordance with the Second Law.

2. Cockroaches acquire evolutionary advantage by feeding on negative entropy food.
3. Female cockroaches carry their eggs until they are hatched, thereby producing minimal entropy.
4. Mating between cockroaches is a reversible process, involving no change in entropy.
5. When a cockroach's life is snuffed out, its entropy increases — this follows directly from the Second Law. However, when a cockroach dies from natural causes, the entropy change is minimal.
6. Smart cockroaches feed on information, rather than on neg-entropy, thereby reducing the entropy of the universe.

References and Suggested Reading

Anfinsen, C.B. (1973), *Science*, **181**, 223.
Atkins, P. (1984), *The Second Law*. Scientific American Books, W. H. Freeman and Co., New York.
Atkins, P. (2007), *Four Laws That Drive the Universe*. Oxford University Press, Oxford, UK.
Baierlein, R. (1994), Entropy and the second law: A pedagogical alternative. *Am J Phys*, **62**, 15–26.
Battino, R. (2007), "Mysteries" of the first and second laws of thermodynamics. *J Chem Educ*, **84**, 753–755.
Ben-Naim, A. (1987), Is mixing a thermodynamic process? *Am J Phys*, **55**, 725–733.
Ben-Naim, A. (1992), *Statistical Thermodynamics for Chemists and Biochemists*. Plenum Press, New York.
Ben-Naim, A. (2006a), *A Molecular Theory of Solutions*. Oxford University Press, Oxford, UK.
Ben-Naim, A. (2006b), The entropy of mixing and assimilation: An information-theoretical perspective. *Am J Phys*, **74**, 1126–1135.
Ben-Naim, A. (2007), *Entropy Demystified: The Second Law of Thermodynamics Reduced to Plain Common Sense*. World Scientific, Singapore.
Ben-Naim, A. (2008), *A Farewell to Entropy: Statistical Thermodynamics Based on Information*. World Scientific, Singapore.
Ben-Naim, A. (2009), An informational-theoretical formulation of the Second Law of thermodynamics. *J Chem Educ*, **86**, 99–105.
Ben-Naim, A. (2010), *Discover Entropy and the Second Law of Thermodynamics: A Playful Way of Discovering a Law of Nature*. World Scientific, Singapore.

Ben-Naim (2011a), *Molecular Theory of Water and Aqueous Solutions. Part II: The Role of Water in Protein Folding, Self-assembly and Molecular Recognition*. World Scientific, Singapore.
Ben-Naim, A. (2011b), Entropy: Order or information. *J Chem Educ*, 88, 594–596.
Ben-Naim, A. (2012), *Entropy and the Second Law: Interpretation and Misss-Interpretationsss*. World Scientific, Singapore.
Ben-Naim, A. (2013), *The Protein Folding Problem and Its Solutions*. World Scientific, Singapore.
Ben-Naim, A. (2015a), *Information, Entropy, Life and the Universe: What We Know and What We Do Not Know*. World Scientific, Singapore.
Ben-Naim, A. (2015b), *Discover Probability: How to Use It, How to Avoid Misusing It, and How It Affects Every Aspect of Your Life*. World Scientific, Singapore.
Ben-Naim, A. (2016a), *The Briefest History of Time*. World Scientific, Singapore.
Ben-Naim, A. (2016b), *Myths and Verities in Protein Folding Theories*. World Scientific, Singapore.
Bent, H.A. (1965), *The Second Law*. Oxford University Press, New York.
Boltzmann, L. (1877), *Vienna Academy*. 42, "*Gesammelte Werke*" p. 193.
Boltzmann, L. (1896), *Lectures on Gas Theory*. Translated by S.G. Brush, Dover, New York (1995).
Bricmont, J. (1996), *Science of Chaos or Chaos of Science*. arXiv:chao-dyn/9603009 V1 22 iyar 1996.
Brillouin, L. (1962), *Science and Information Theory*. Academy Press, New York.
Brush, S.G. (1976), *The Kind Of Motion We Call Heat: A History Of The Kinetic Theory of Gases In The 19th Century. Book 2: Statistical Physics and Irreversible Processes*. North Holland Publishing Company.
Brush, S.G. (1983), *Statistical Physics and the Atomic Theory of Matter, from Boyle and Newton to Landau and Onsager*. Princeton University Press, Princeton.
Callen, H.B. (1960), *Thermodynamics*. John Wiley and Sons, New York.
Callen, H.B. (1985), *Thermodynamics and an Introduction to Thermostatics*, 2nd Edition. John Wiley and Sons, New York.
Cooper, L.N. (1968), *An Introduction to the Meaning and Structure of Physics*. Harper and Row, New York.
Cover, T.M. and Thomas, J.A. (1991), *Elements of Information Theory*. John Wiley and Sons, New York.

Denbigh, K. (1981), How subjective is entropy? *Chem Br*, **17**, 168–185.
Denbigh, K.G. and Denbigh, J.S. (1985), *Entropy in Relation to Incomplete Knowledge*. Cambridge University Press, Cambridge.
Denbigh, K.G. (1989), Note on entropy, disorder and disorganization. *Br J Philos Sci*, **40**, 323–331.
Dugdale, J.S. (1996), *Entropy and its Physical Meaning*. Taylor and Francis, London.
Eddington, A. (1928), *The Nature of the Physical World*. Cambridge University Press, Cambridge, UK.
Eisenberg, D. and Kauzmann, W. (1969), *The Structure and Properties of Water*. Oxford University Press, Oxford, UK.
Eliel, E.L. (1962), *Stereochemistry of Carbon Compounds*. McGraw-Hill Book Company, Inc., New York.
Eley, D.D. (1939), On the solubility of gases. Part 1: The inert gases in water. *Trans Faraday Soc*, **35**, 1281–1293.
Eley, D.D. (1944), *Trans. Faraday. Soc.* **40**, 184.
Fast, J.D. (1962), *Entropy. The Significance of the Concept of Entropy and Its Applications in Science and Technology*. Philips Technical Library, Netherlands.
Frank, H.S. and Evans, M.W. (1945), Free volume and entropy in condensed systems. III. Entropy in binary liquid mixtures; partial molal entropy in dilute solutions; structure and thermodynamics in aqueous electrolytes. *J Chem Phys*, **13**, 507–532.
Gamow, G. (1940), *Mr. Tompkins in Wonderland*. Cambridge University Press, Cambridge, UK.
Gamow, G. and Stannard, R. (1999), *The New World of Mr. Tompkins*. Cambridge University Press, Cambridge, UK.
Gell-Mann, M. (1994), *The Quark and the Jaguar*. Little Brown, London.
Gibbs, J.W. (1906a), *Collected Scientific Papers of J. Willard Gibbs*. Longmans, Green, and Co., New York.
Goldsein, S. (2001), *Boltzmann's Approach to Statistical Mechanics*. (Published in arXiv:cond-mat/0105242,v1, 11 May 2001).
Goldstein, H., Poole, C.P., and Safko, J.L. (2002), *Classical Mechanics*, 3rd Edition. Addison, Wesley, New York.
Greene, B. (2004), *The Fabric of the Cosmos: Space, Time, and the Texture of Reality*. Alfred A. Knopf.
Guggenheim, E.A. (1949), *Statistical Basis of Thermodynamics*. Research, **2**, 450–454.
Hawking, S.W. (1988), *A Brief History of Time: From the Big Bang Theory to Black Holes*. Bantam Books, New York.

Hill, T.L. (1960), *Introduction to Statistical Mechanics*. Addison-Wesley, Reading, Massachusetts.
Hoffman, P.M. (2012), *Life's Ratchet: How Molecular Machines Extract Order from Chaos*. Basic Books, New York.
Jaynes, E.T. (1957a), Information theory and statistical mechanics. *Phys Rev*, 106, 620.
Jaynes, E.T. (1957b), Information theory and statistical mechanics II. *Phys Rev*, 108, 171.
Jaynes, E.T. (1965), Gibbs vs Boltzmann entropies. *Am J Phys*, 33, 391–398.
Kafri, O. and Kafri H. (2013), *Entropy: God's Dice Game*. CreateSpace Independent Publishing Platform.
Katchalsky, A. (1963), Nonequilibrium thermodynamics. *International Science and Technology*, 43–49.
Katz, A. (1967), *Principles of Statistical Mechanics: The Informational Theory Approach*. W.H. Freeman, London.
Kauzmann, W. (1959), Some factors in the interpretation of protein denaturation. *Adv Protein Chem*, 14, 1–63.
Khinchin, A.I. (1957), *Mathematical Foundation of Information Theory*. Dover, New York.
Kozliak, E.I. and Lambert, F.L. (2005), "Order-to-disorder" for entropy change? Consider the numbers! *Chem Educ*, 10, 24–25.
Lambert, F.L. (1999), Shuffled cards, messy desks and disorderly dorm rooms: Examples of entropy increase? Nonsense! *J Chem Educ*, 76, 1385–1387.
Lambert, F.L. (2002), Entropy is simple, qualitatively. *J Chem Educ*, 79, 1241–1246.
Lambert, F.L. (2007), Configurational entropy revisited. *J Chem Educ*, 84, 1548–1550.
Lebowitz, J.L. (1993), Boltzmann's entropy and time's arrow. *Phys Today*, 46, 32–38.
Lebowitz, J.L. (1999), Microscopic origins of irreversible macroscopic behavior. *Physica A*, 263, 516–527.
Leff, H.S. (1966), Thermodynamics entropy: The spreading and sharing of energy. *Am J Phys*, 64, 1261–1267.
Lemons, D.S. (2013), *A Student's Guide to Entropy*. Cambridge University Press, Cambridge, UK.
Lewis, G.N. (1930), The symmetry of time in physics. *Science*, 71, 569–577.
Lindley, D.V. (1965), *Introduction to Probability and Statistics*. Cambridge University Press, Cambridge, UK.
Mackey, M.E. (2003), *Time's Arrows: The Origin of Thermodynamic Behavior*. Dover, New York.

Mayhew, K.W. (2015a), *Changing Our Perspective, Part I: A New Thermodynamics*. CreateSpace Independent Publishing Platform.
Mayhew, K.W. (2015b), An ill-conceived mathematical contrivance. *Physics Essays*, **28**, 358.
Nordholm, S. (1997), In defense of thermodynamics: An animate analogy. *J Chem Educ*, **74**, 273–275.
Pauling, L. (1939), *The Nature of the Chemical Bond*, 2nd Edition. Cornell University Press, Ithaca, New York.
Penrose, R. (1989), *The Emperor's New Mind*. Oxford University Press, Oxford, UK.
Prigigine, I. (1997), *The End of Certainty, Time Chaos and the New Laws of Physics*. The Free Press, New York.
Rifkin, J. (1980), *Entropy: A New World View*. Viking Press, New York.
Rushbrooke, G.S. (1949), *Introduction to Statistical Mechanics*. Clarendon Press, Oxford, UK.
Sackur, O. (1911), The application of the kinetic theory of gases to chemical problems. *Annalen der Physik*, **36**, 958–980.
Sanford, J.C. (2005), *Genetic Entropy and the Mystery of the Genome*. Ivan Press, a division of Elim Publishing, New York.
Scully, R.J. (2007), *The Demon and the Quantum. From the Pythagorean Mystics to Maxwell's Demon and Quantum Mystery*. Wiley-VCH Verlag GmbH & Co. kGaA.
Shannon, C.E. (1948), A mathematical theory of communication. *Bell System Tech J*, **27**, 379–423.
Sheehan, D.P. and Gross, D.H.E. (2006), Extensivity and the thermodynamic limit: Why size really does matter. *Physica A*, **370**, 461–482.
Sommerfeld, A. (1956), *Thermodynamics and Statistical Mechanics*. Academic Press, New York.
Styer, D.F. (2000), Insight into entropy. *Am J Phys*, **68**, 1090–1096.
Styer, D.F. (2008), Entropy and evolution. *Am J Phys*, **76**, 1031–1033.
Tetrode, H. (1912), The chemical constant of gases and the elementary quantum of action. *Annalen der Physik*, **38**, 434–442.
Thomson, W. (1874), The kinetic theory of the dissipation of energy. *Proceeding of the Royal Society of Edinburgh*, **8**, 325–334.
Tribus, M. and McIrvine, E.C. (1971), Entropy and information. *Scientific American*, **225**, 179–188.
Volkenstein, M.V. (2009), *Entropy and Information*. Birkhäuser, Berlin.
Von Baeyer, H.C. (2005), *Information: The New Language of Science*. Harvard University Press, Cambridge, Massachusetts.
Wilks, J. (1961), *The Third Law of Thermodynamics*. Oxford University Press, Oxford, UK.

Index

Arrow of time, 223–225

Chemical equilibrium, 181–183
Clapeyron equation, 188

Entropy
    absolute value, 25–27, 160–165
    and information, 135–140
    and disorder, 120–129
    and freedom, 148–149
    and spreading, 129–135
    and the arrow of time, 223–225
    application to life, 115, 216–219
    as unavailable energy, 152–153
    as uncertainty, 155–156
    definition, based on SMI, 27–58
    definition, Boltzmann, 23–25, 22–27
    definition, Clausius, 5–22
    measure of possibilities, 149–150
    measure of irreversibility, 150–152
    mystery of, 105–113
    of assimilation, 202–206
    of expansion, 17–22
    of mixing, 10, 199–206
    of mixtures of enantiomers, 153–154
    of solvation, 193–199
    of the universe, 219–223
    subjective, 140–148
Entropy change in expansion process, 17–22

Heat engine, 7–10
Hydrogen formation, 175–178

Phase transition, 187–191
    of water, 188–191
Protein folding, 183–187
Protein-protein association, 178–181

Racemization, 206–214
Residual entropy, 160–165
  of ice, 165–173
Reversal paradox, 113–115

Shannon's measure of information
  (SMI), 27–41
  and entropy, 45–58
  interpretations of, 41–45
Solubility of argon, 215–216
Solvation of argon, 193–199

Trouton Law, 191–192
The Second Law
  and life, 216–219
  entropy formulation, 58–93, 175–177
  The T, P, N formulation, 94–96, 174–178
  The T, V, N formulation, 93, 174
Twenty question (20Q) game, 32–41

World Scientific
Connecting Great Minds

Other Recent Books by the Author

The Briefest History of Time
The History of Histories of Time and the Misconstrued Association between Entropy and Time

By: Arieh Ben-Naim

ISBN: 978-981-4749-84-8
ISBN: 978-981-4749-85-5 (pbk)

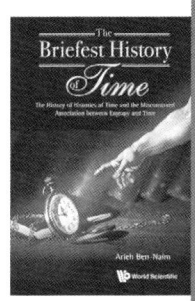

Entropy Demystified
The Second Law Reduced to Plain Common Sense
2nd Edition

By: Arieh Ben-Naim

ISBN: 978-981-3100-11-4
ISBN: 978-981-3100-12-1 (pbk)

Myths and Verities in Protein Folding Theories

By: Arieh Ben-Naim

ISBN: 978-981-4725-98-9
ISBN: 978-981-4725-99-6 (pbk)

Information, Entropy, Life and the Universe
What We Know and What We Do Not Know

By: Arieh Ben-Naim

ISBN: 978-981-4651-66-0
ISBN: 978-981-4651-67-7 (pbk)

World Scientific
Connecting Great Minds

Statistical Thermodynamics
With Applications to
the Life Sciences

By: Arieh Ben-Naim

ISBN: 978-981-4579-15-5
ISBN: 978-981-4578-20-2 (pbk)

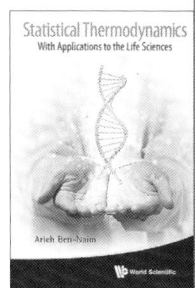

The Protein Folding Problem and Its Solutions

By: Arieh Ben-Naim

ISBN: 978-981-4436-35-9
ISBN: 978-981-4436-36-6 (pbk)

Entropy and the Second Law
Interpretation and
Misss-Interpretationsss

By: Arieh Ben-Naim

ISBN: 978-981-4407-55-7
ISBN: 978-981-4374-89-7 (pbk)

Discover Entropy and the Second Law of Thermodynamics
A Playful Way of Discovering
a Law of Nature

By: Arieh Ben-Naim

ISBN: 978-981-4299-75-6
ISBN: 978-981-4299-76-3 (pbk)